International Zinc and Zinc Oxide Conferences 2020

Scottsdale, Arizona, USA
16 - 19 February 2020

ISBN: 978-1-7138-0807-7

Printed from e-media with permission by:

Curran Associates, Inc.
57 Morehouse Lane
Red Hook, NY 12571

Some format issues inherent in the e-media version may also appear in this print version.

International Zinc Association
2530 Meridian Parkway
Suite 115
Durham, NC 27713
USA

Phone: +1 919 361 4647

contact@zinc.org

Additional copies of this publication are available from:

Curran Associates, Inc.
57 Morehouse Lane
Red Hook, NY 12571 USA
Phone: 845-758-0400
Fax: 845-758-2633
Email: curran@proceedings.com
Web: www.proceedings.com

TABLE OF CONTENTS

Statement of Compliance

1. **2020 International Zinc & Zinc Oxide Conferences, Scottsdale, AZ, February 16-19, 2020.**
2. The purpose of the meeting was to address issues concerning zinc producers; to promote the public awareness of and interest in the zinc industry; and to monitor governmental activities and related tasks.
3. No discussions were held, formally or informally, during specified meeting times or otherwise, involving, directly or indirectly, express or implicit agreements or understandings related to: (a) any company's price; (b) any company's terms or conditions of sale; (c) any company's production or sales levels; (d) any company's wages or salaries; (e) the division or allocation of customers or geographic markets; or (f) customer or suppliers boycotts.
4. The International Zinc Association as a group makes no recommendations of any kind and has not reached any agreements or understandings with respect to an individual company's prices, terms or conditions of sale, production or sales levels, wages, salaries, customers or suppliers.
5. I have read the foregoing and it is true to the best of my knowledge and belief. This statement is freely given without threat or reprisal of promise of benefit.

______________________ Name	______________________ Title
______________________ Company	______________________ Date

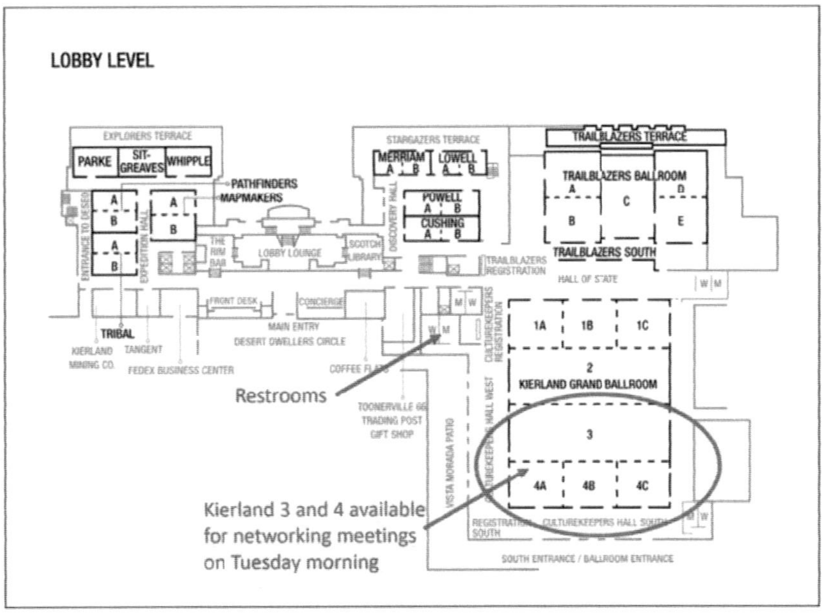

SPONSORS

GLENCORE

CLIVEDEN TRADING AG

PEÑOLES

ARTAZN

Richker Metals Inc.

TRAXYS

PGS

ocean PARTNERS

Iceberg – what lies beneath?

– today's zinc market and outlook

International Zinc Conference
Scottsdale – 18th Feb 2020

Concord Resources Limited

Duncan Hobbs – Research Director

duncan.hobbs@concordltd.com

CONCORD

Disclaimer

Concord Resources Limited
Who we are

- **Founded in 2015** Concord is a global resources merchant focused on the supply chain, financing and risk management of **non-ferrous metals and associated raw materials**, led by an experienced and entrepreneurial management team
- Concord is a global company, operating from its **headquarters in London** and through its principal **offices in New York and Hong Kong**. Our experienced team work with resource producers and consumers on spot transactions out to multi-year, structured agreements
- Concord's capabilities include **global sourcing, marketing and management of all the required logistics and finance** to optimise our clients' supply chains from long-term partnership marketing arrangements to single container deliveries
- Concord brings substantial cross-commodity, structuring and asset optimisation experience of its founding team to work for the **benefit of its customers and partners**
- Concord is an **independent company**, and combines its founding management's experience with the commodity and financial industry expertise of its shareholders
- For more information please visit our website

CONCORD 2

Iceberg – what lies beneath?
Today's zinc market and outlook

CONCORD 3

Iceberg – what lies beneath?
Today's zinc market and outlook

Context

LME zinc stocks recently at lowest levels in over 30 years

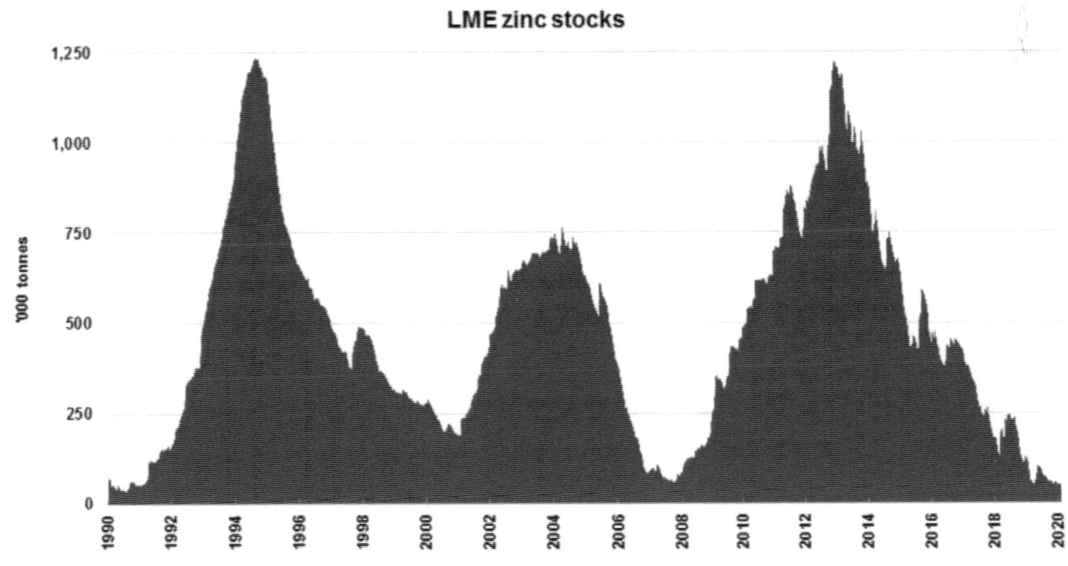

Note: Data to 13 Feb 2020. Since April 1997, LME stock data have been reported daily; previously stock data were reported only twice weekly
Source: LME, Concord Resources Limited, Feb 2020

Context

SHFE zinc stocks recently near lowest levels since contract launched in 2007

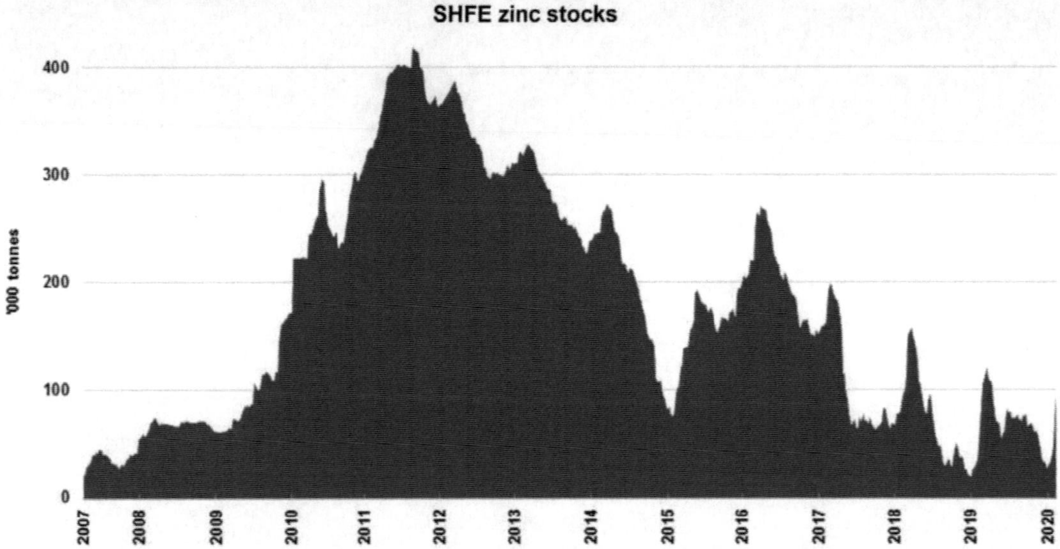

Note: SHFE zinc contract started trading in Apr 2007. Data to 14 Feb 2020
Source: SHFE, Concord Resources Limited, Feb 2020

Context

Decline in exchange stocks even more pronounced vs zinc market size

Note: Calculations basis month-end stock data or latest available vs monthly consumption data annualised
Source: ILZSG, LME, SHFE, Concord Resources Limited, Feb 2020

Context

Most stats show zinc metal market running significant deficit in recent years

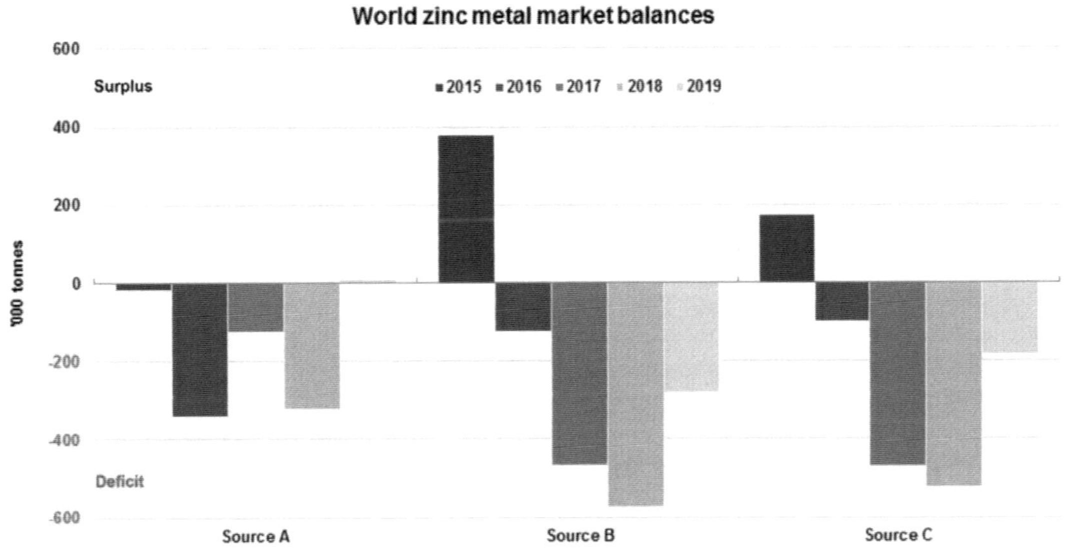

Note: 2019 data include some estimates
Source: Concord Resources Limited, Feb 2020

Context

Extended backwardation seen in LME zinc spreads over last two years

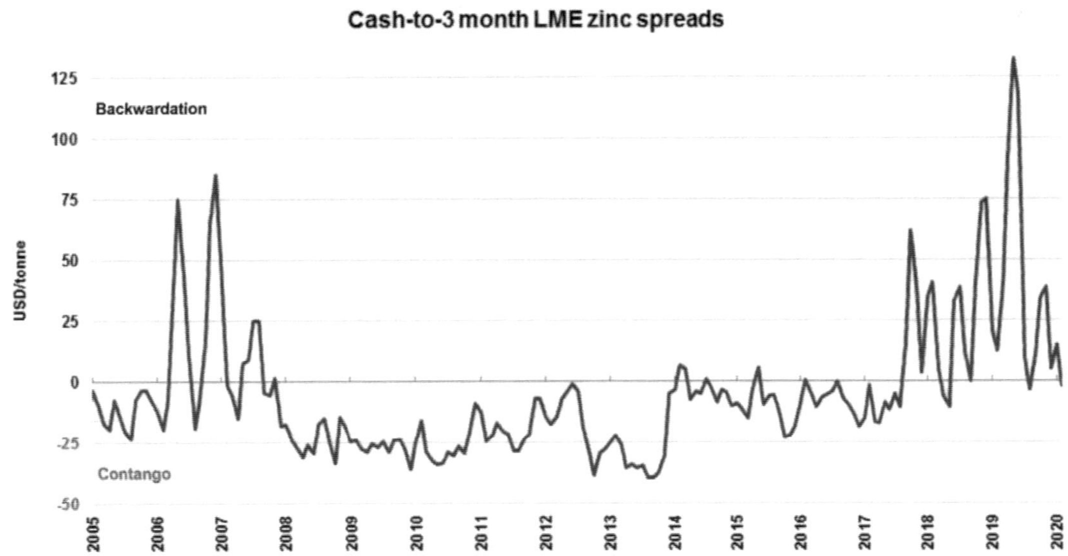

Note: Month averages. Data to 14 Feb 2020
Source: LME, Concord Resources Limited, Feb 2020

Context

LME-SHFE zinc arb negative for many months now

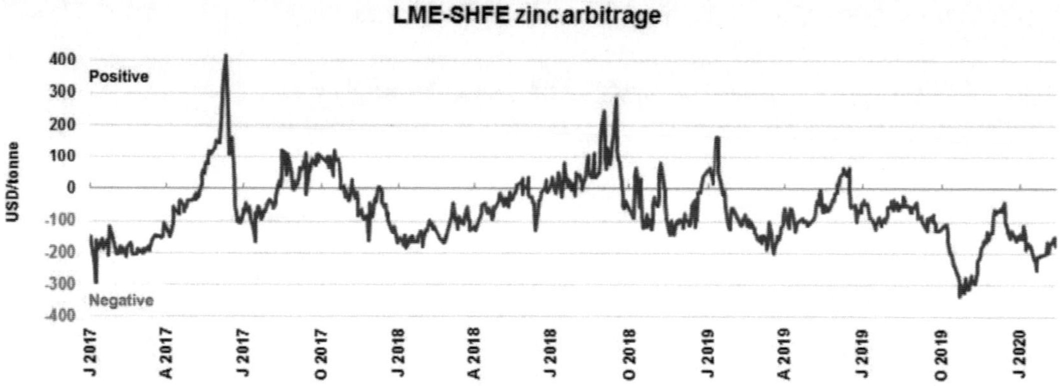

- China is structurally a net import market for zinc metal, averaging 500ktpa since the mid-2000s
- China's zinc metal imports fell -16% in 2019 to 685kt; 3-year low

Note: Data to 14 Feb 2020
Source: Fastmarkets, LME, SHFE, Concord Resources Limited, Feb 2020

Context

Zinc metal premiums have dropped to multi-year lows in most markets

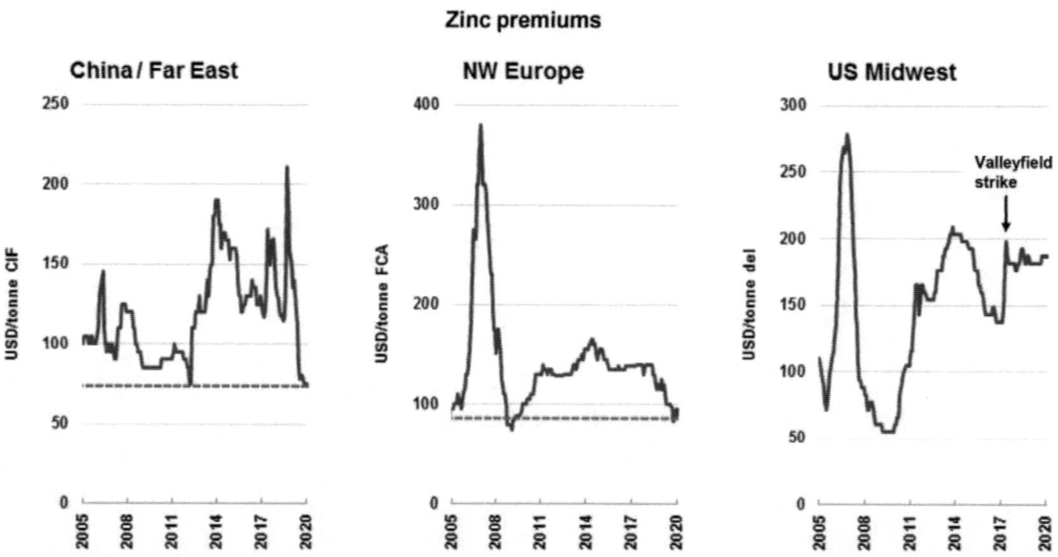

Note: Spot SHG ingot premiums
Source: CRU, Fastmarkets, Concord Resources Limited, Feb 2020

Context

Hard data point to fall in real demand for zinc metal in 2018-2019

Output levels in key markets for zinc metal

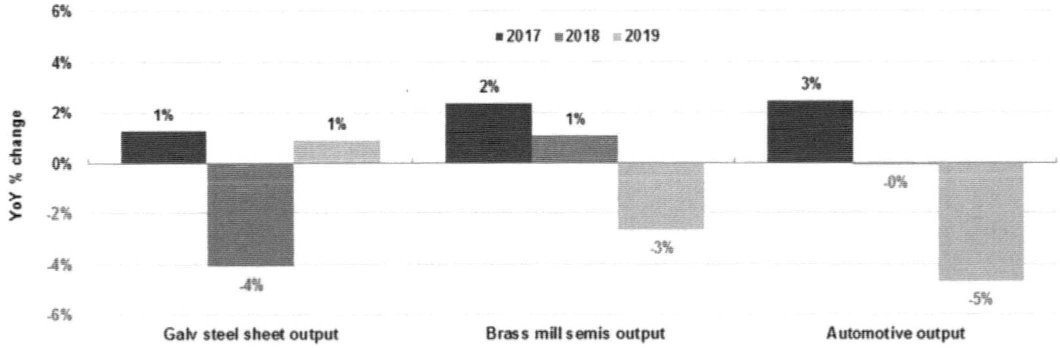

- **Galvanised steel production (±50%) and brass making (±17%) are key markets for zinc consumption**
- **Automotive output is a key market for galvanised steel and zinc die-cast alloys**

Note: 2019 data include estimates. Automotive output data include cars and light commercial vehicles only
Source: CRU, ILZSG, Concord Resources Limited, Feb 2020

CONCORD 12

Context

Structural change curbing zinc demand in key markets – continuous galvanising

EU market for ZAM / ZM coated steel sheet

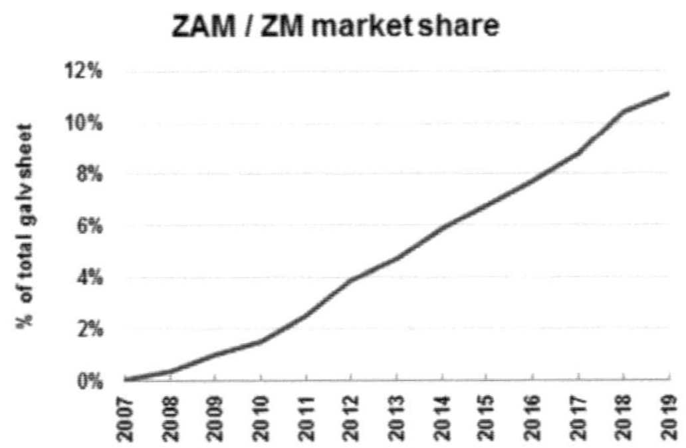

- **ZAM / ZM coatings applied in continuous galvanising 20%-40% lighter than free-zinc – SteelConsult**
- **N.American & Asian steelmakers are also promoting ZAM / ZM coatings**

Note: ZAM – Zinc-Aluminium-Magnesium. ZM – Zinc-Magnesium. 2019 data are estimates
Source: SteelConsult *(ZnMg Coated Steel Sheet in the EU)*, Concord Resources Limited, Feb 2020

CONCORD 13

Context

Structural change curbing zinc demand in key markets – batch galvanising

Innovation path for zinc coatings (HDG)

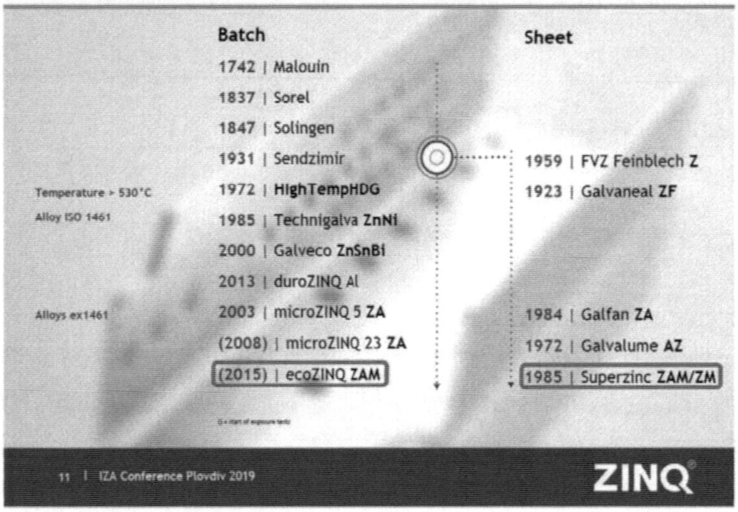

- **ZAM / ZM coatings were introduced in continuous galvanising 30 years ago**
- **Leading batch galvanisers have started introducing ZAM / ZM coatings only around five years ago**
- **ZINQ ranks 2nd in the world league of batch hot-dipped galvanisers based on revenues**

Source: ZINQ, Concord Resources Limited, Feb 2020

Context

Estimates of cumulative zinc metal market balance differ over last decade

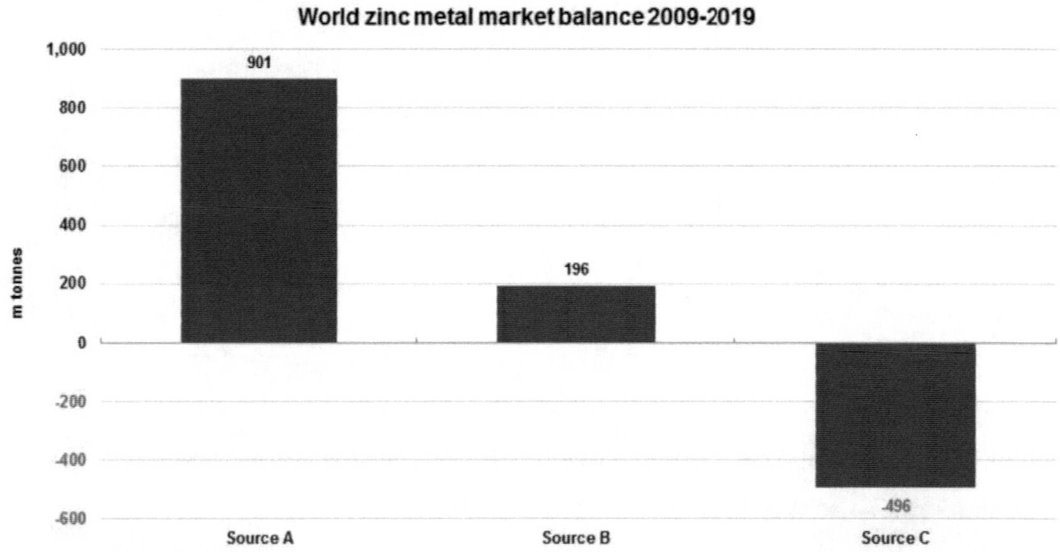

Note: 2019 data include some estimates
Source: Concord Resources Limited, Feb 2020

Context

Squaring the circle: statistical picture vs physical market experience

Zinc consumption

- Calculating 'consumption' of zinc (any metal) is very difficult; most data are estimates only
  - There are very few stats on real consumption of zinc metal
  - Stock data needed to translate 'apparent' to 'real' consumption are largely lacking
- Real zinc consumption data may have been overstated – consider, for example:
  - ZAM / ZM coatings reported gaining market share in galvanising
  - Reported decline of DCAs market

Zinc stock levels and holdings

- Some sources may have understated zinc metal market surpluses in the fall out from the GFC
- More stocks may have been built in practice than on paper, helping to buffer later metal market deficits
- Post-GFC QE / cheap money, change in industry structure may have impacted where, how much stock held

Source: Concord Resources Limited, Feb 2020

Iceberg – what lies beneath?

Today's zinc market and outlook

Market outlook

World zinc consumption is closely correlated with industrial production

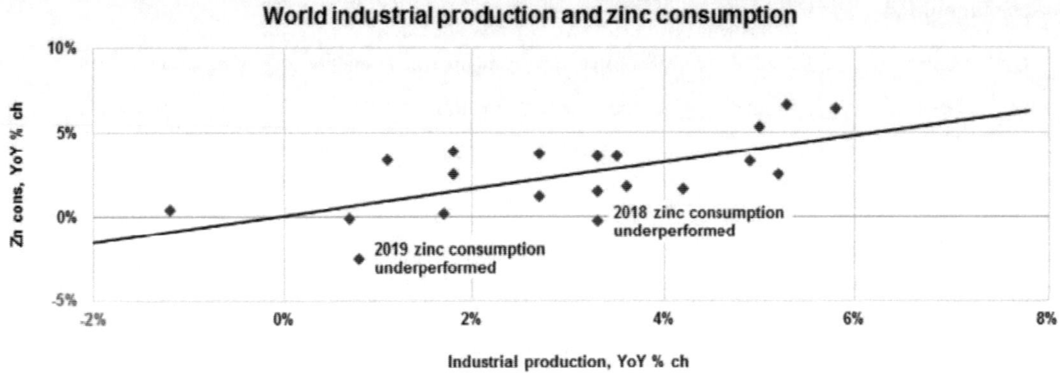

- **YoY change in industrial production 'explains' majority of YoY change in world zinc consumption**
- **Note negative intercept on y-axis – i.e. zinc consumption typically lags IP**

Note: Data from 2000-2019 (exc. 2009-2010 (GFC dislocation)). 2019 data are estimates
Source: CPB, CRU, Concord Resources Limited, Feb 2020

Market outlook

Latest data point to slower growth in IP and, by association, zinc consumption

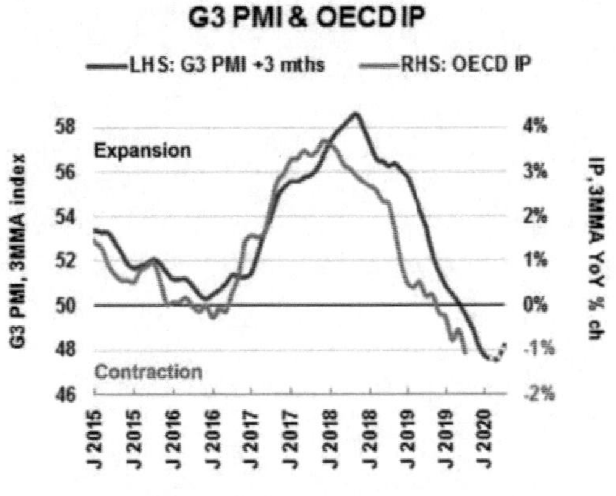

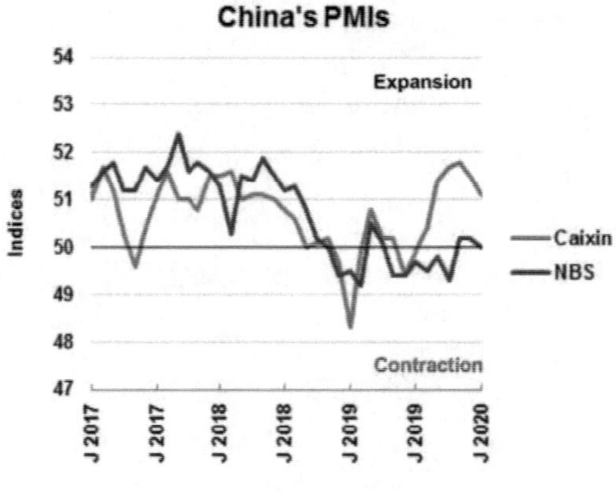

- **3MMA G3 Feb PMI < 50 for 7th month running in Jan; up +0.6 pts MoM to 48.2 (before COVID-19)**
- **China's Jan NBS PM down -0.2 pts MoM to 50.0; Markit PMI also down to 51.1, 5-month low**

Note: PMIs for manufacturing. G3 – USA, Eurozone & Japan. G3 PMI calculation weighted by industrial production
Source: ISM, Markit, NBS, OECD, Concord Resources Limited, Feb 2020

Market outlook

Zinc consumption forecast to rise in 2020, but still well below long-run rate

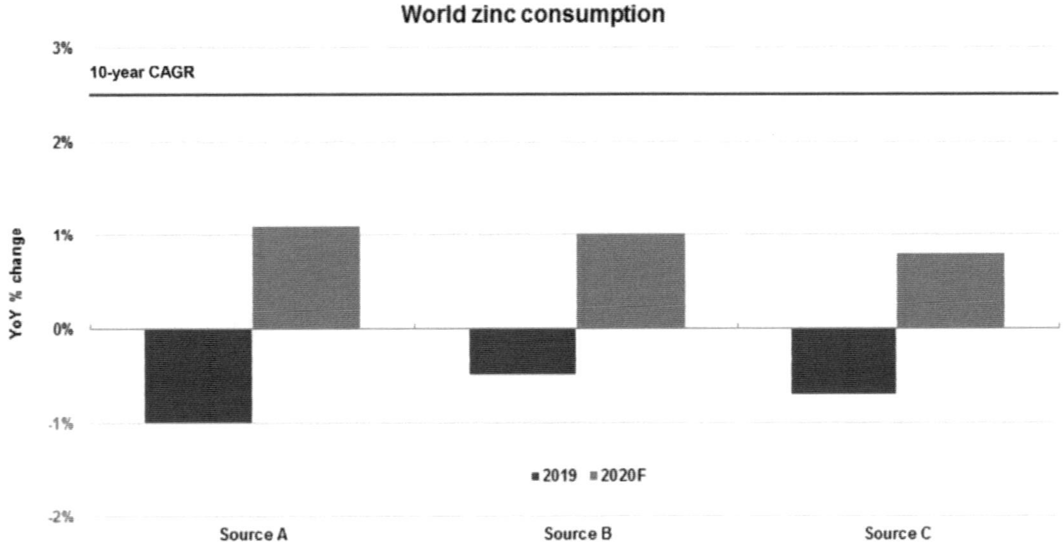

Note: 10-year CAGR is an average of Sources A, B & C
Source: Concord Resources Limited, Feb 2020

Market outlook

QE – keep an eye on the once more rising tide

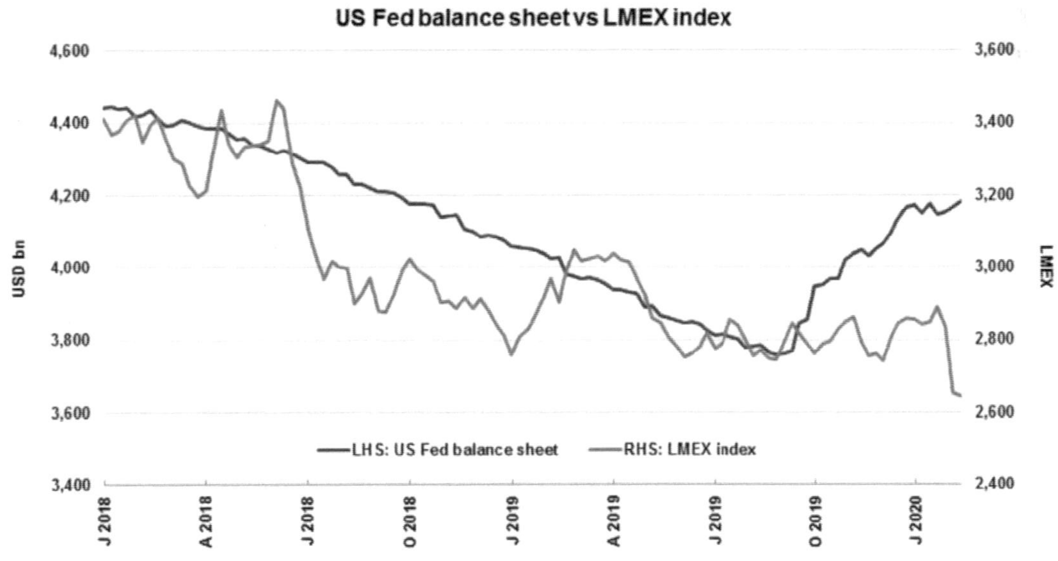

Note: Data to 12 Feb 2020
Source: Concord Resources Limited, Feb 2020

Market outlook

Overcooked – US dollar looks expensive on this menu!

The Economist's 'Big Mac' index

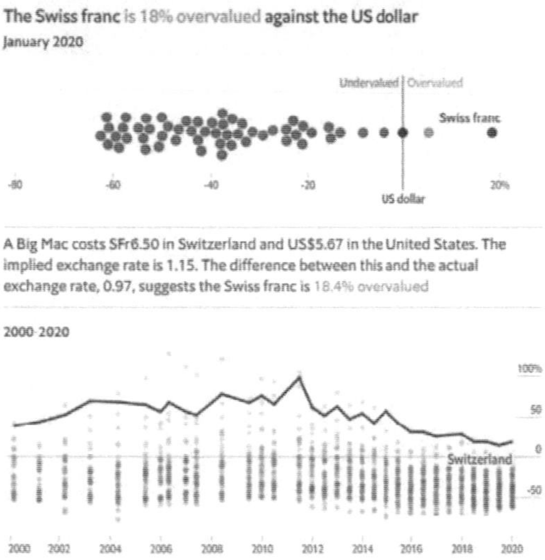

US dollar trade-weighted index

Note: US dollar index month average data to 7 Feb 2020
Source: The Economist, US Federal Reserve, Concord Resources Limited, Feb 2020

Market outlook

Zinc smelters realising very high unit revenues today

Zinc smelter revenues

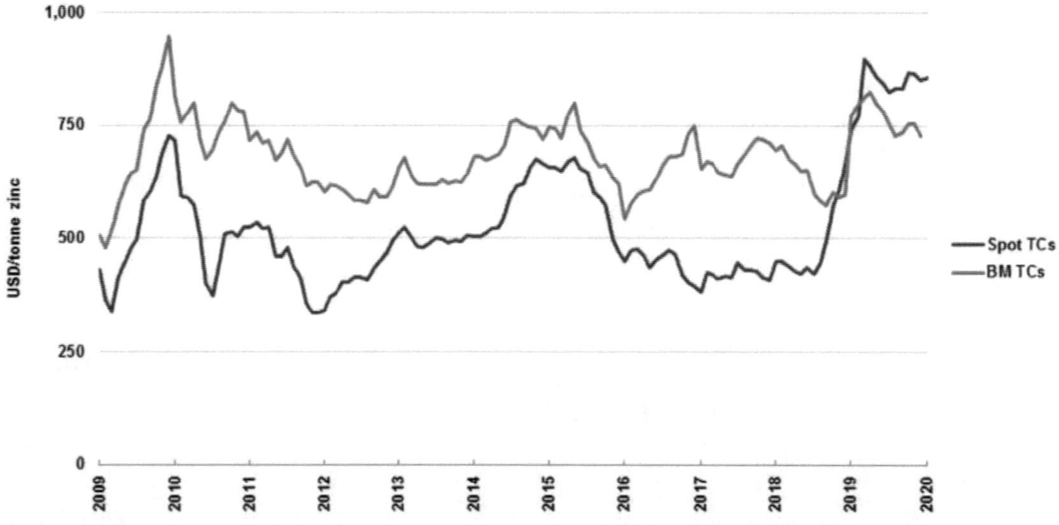

Note: TCs for 'clean' zinc concs on standard terms. 2020 benchmark had not been reported at the time of publication. Revenues excluding by-products and metal premiums
Source: CRU, Fastmarkets, LME, Concord Resources Limited, Feb 2020

Market outlook

China's zinc mine output up despite sharp drop in unit revenue in 2019

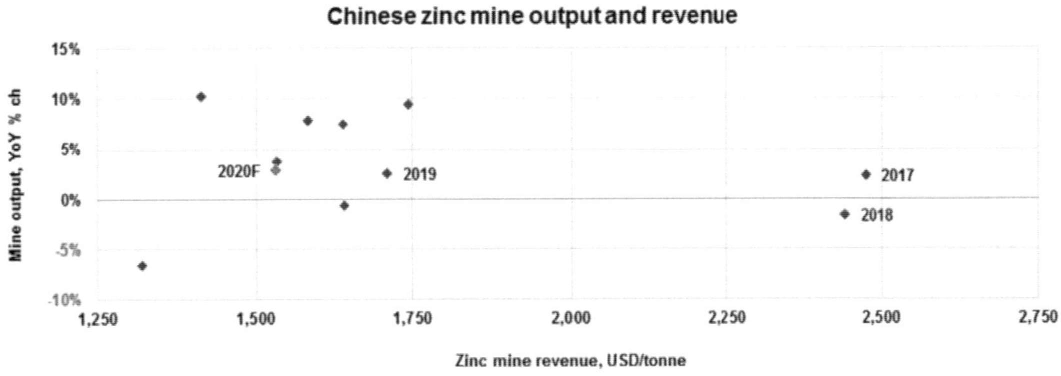

- **Relationship between mine revenue and output growth broke down in 2018-2019, recovered in 2019**
- **2020 mine output growth forecast in range of past performance basis revenues in January**

Note: 2020 data point: mine output forecast, mine revenue basis average prices and TCs for Jan only
Source: CRU, FM, LME, Concord Resources Limited, Feb 2020

CONCORD

Context

Ex-China zinc mine output rising strongly, more on the way

Key mine expansions in 2020

- Antamina +145kt
- Gamsberg, Vedanta +95kt
- Penasquito, Newmont +75kt
- Rampura, Vedanta +75kt
- New Century +65kt
- Sindesar Khurd, Vedanta +65kt
- Zhairem, Glencore +45kt
- Zawar Mines, Vedanta +40kt
- Iscaycruz, Glencore +40kt
- Neves Corvo, Lundin +30kt

Note: Ex-China zinc mine output data to Nov 2019
Source: Company Reports, CRU, ILZSG, Concord Resources Limited, Feb 2020

CONCORD

Market outlook

Zinc mines' unit revenues have fallen below long-run average

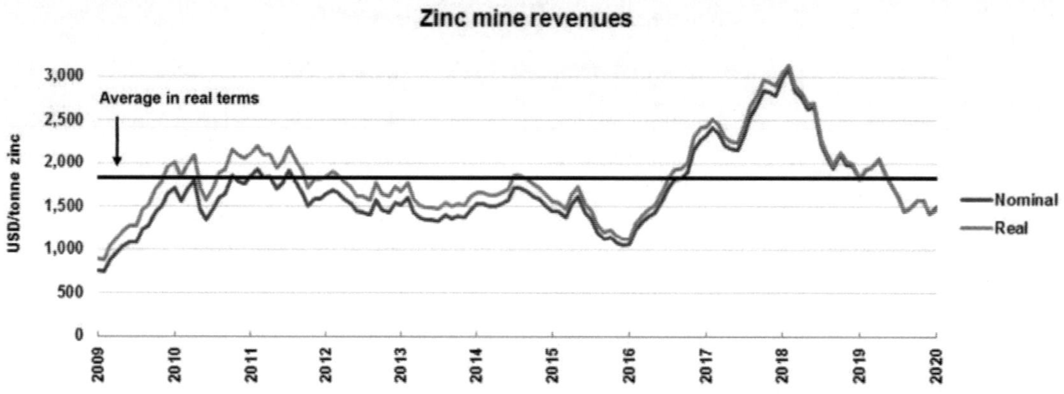

- **Any further sharp / sustained falls in zinc price could put forecast rise in mine output at risk**
- **Higher TCs (i.e. higher mine costs) risk the same impact (unless offset by higher prices)**

Note: Basis spot TCs for 'clean' zinc concs on standard terms, excluding by-products. Real terms data calculated basis US GDP deflator 2019 = 100
Source: CRU, Fastmarkets, LME, Concord Resources Limited, Feb 2020

Market outlook

Zinc TCs have been about flat for months, despite increasingly negative arb

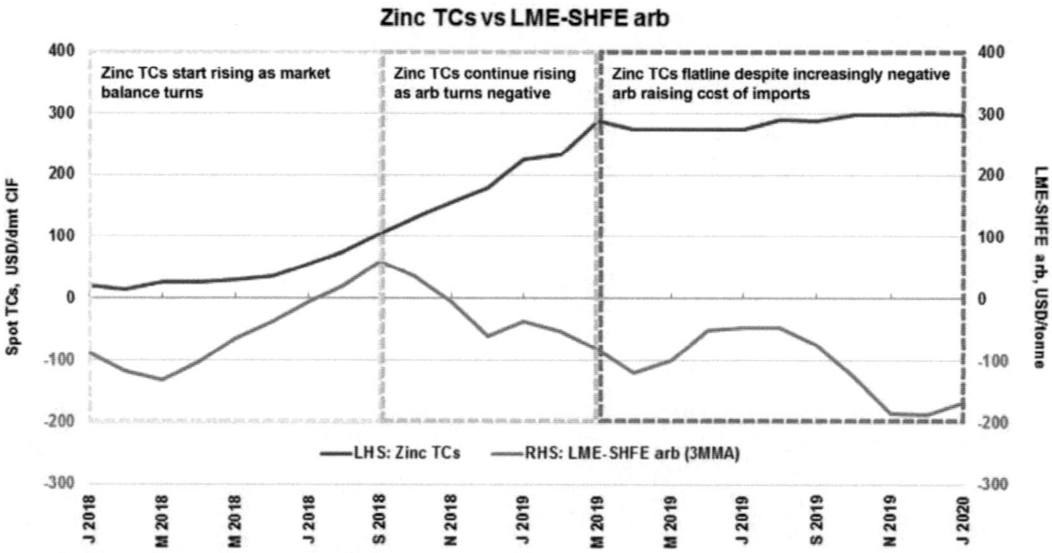

Note: TCs for 'clean' zinc concs on standard terms
Source: CRU, Fastmarkets, LME, SHFE, Concord Resources Limited, Feb 2020

Iceberg – what lies beneath?
Today's zinc market and outlook

Summary
Iceberg – what lies beneath? Today's zinc market and outlook

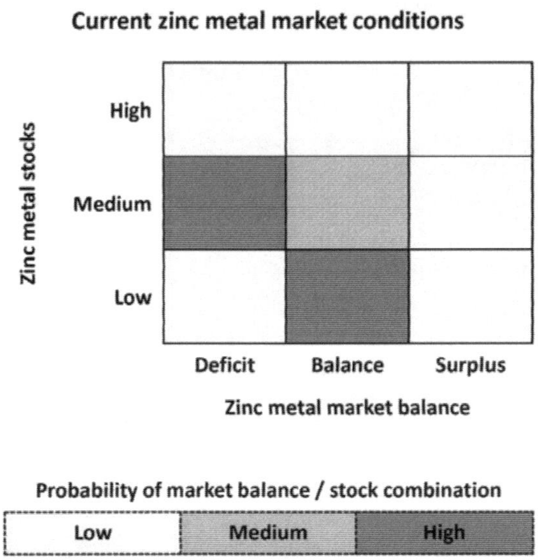

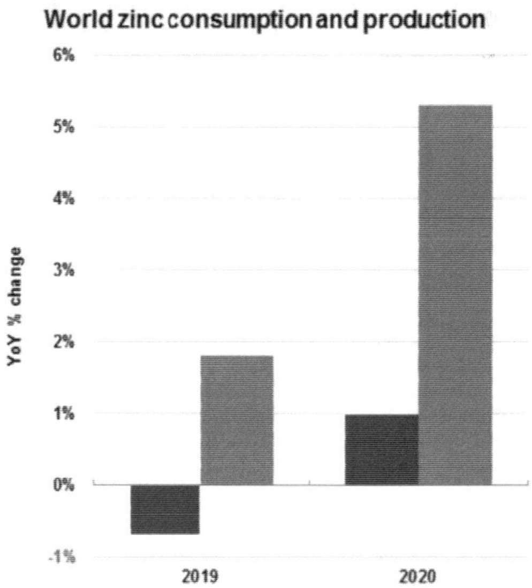

Note: Average of world zinc consumption and production data from selected research firms. 2020 data are forecasts
Source: Concord Resources Limited, Feb 2020

Summary

Iceberg – what lies beneath? Today's zinc market and outlook

Today's zinc market

- Paradox between strong statistical picture vs soft physical market experience
  - Metal market deficits shown on paper, low exchange stocks, tight time spreads
  - China arb negative, falling / low premiums, negative demand data / structural change
- Recent metal market deficits maybe overstated
- Stock build following GFC maybe understated
- Stock holding dynamics may have changed
  - QE / cheap money, market consolidation

Zinc market outlook for 2020

- Leading economic indicators had been improving – pre-coronavirus escalation; now uncertain
- Forecasts for zinc metal consumption up YoY, but < 50% of long run average
- Zinc smelter revenues are buoyant – strong financial incentive to push metal production
- Zinc mine output rising fast, *faster than demand*
- Metal market looks set to swing into surplus
- But zinc mine revenues are being squeezed
  - At risk from further price fall / rise in TCs

Source: Concord Resources Limited, Feb 2020

CONCORD 30

Concord Resources Limited

Our markets: every step of the supply chain

	Aluminum	Copper	Zinc, lead, nickel	Alumina, bauxite	Precious metals
Metals				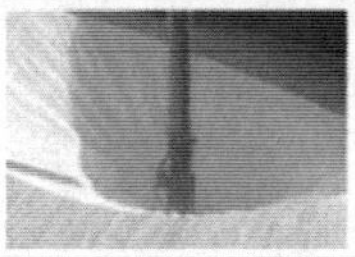	
Products	• P1020 • Billets • Alloys • Rod	• Cathode • Copper rod • Concentrates	• SHG & CGG zinc • 397 & 49s lead • Primary nickel • Concentrates	• SGA • Raw bauxite	• Gold • Silver • PGMs
Origination	• Consignments • Prepays/financing • Off-take/marketing • Tolling	• Consignments • Prepays/financing • Off-take/marketing • Tolling	• Consignments • Prepays/financing • Off-take/marketing • Tolling	• Swaps • Prepays/financing • Off-take/marketing • Tolling	• Off-take/marketing • By-products from concentrates
Storage / logistics	• LME system • Warehousing • Shipping	• LME system • Warehousing • Shipping • Blending	• LME system • Warehousing • Shipping • Blending	• Warehousing • Shipping • Blending	• Vaulting
Consumers	• Aerospace • Automotive • Construction • Packaging	• Wire and cable • Tubing • Brass • Smelters	• Galvanizers • Battery makers • Stainless & alloy steelmakers	• Refineries • Smelters	• Banks • Smelters

CONCORD 31

Concord Resources Limited

Contact our marketing teams

London	New York	Hong Kong
3 ST JAMES'S SQUARE LEVEL 8 LONDON, SW1Y 4JU ENGLAND	437 MADISON AVENUE 28TH FLOOR NEW YORK, NY 10022 USA	41 CONNAUGHT ROAD CENTRAL LEVEL 16 NEXXUS BUILDING HONG KONG
David Freeland Senior Trader EMEA Tel: +44 20 3879 3995 david.freeland@concordltd.com	Jeff Romanek Senior Trader Americas Tel: +1 212 822 9792 jeff.romanek@concordltd.com	Felix Cauro Senior Trader Asia Tel: +852 3757 9915 felix.cauro@concordltd.com
Kareem Barbir Senior Trader Concentrates Tel: +44 20 3879 3993 kareem.barbir@concordltd.com		

Trends in Zinc-Containing Paint Development

Frank E. Goodwin
International Zinc Association

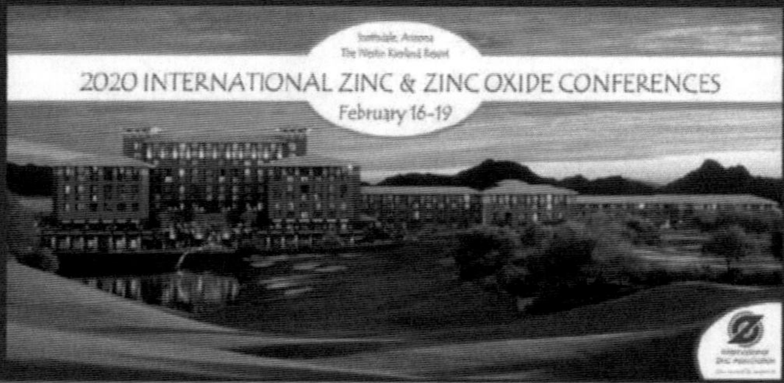

International Zinc Association

Contents

- Introduction
- Past Zn industry work:
  -Required level of Zn in Zn-Rich Paint
  -Improving Stability of Water-based Zn-Containing Paint

Recent Developments
  -Nano Zinc Powder
  -Nano-Fillers
  -Alternative Zn Shapes and Compositions
  -Paint Solvents Development

International Zinc Association

Code of Federal Regulations
Title 40. Protection of Environment
Chapter I. ENVIRONMENTAL PROTECTION AGENCY
Subchapter C. AIR PROGRAMS
Part 59. NATIONAL VOLATILE ORGANIC COMPOUND EMISSION STANDARDS
FOR CONSUMER AND COMMERCIAL PRODUCTS
Subpart D. National Volatile Organic Compound Emission Standards for Architectural Coatings

Definitions:

Metallic pigmented coating means a nonbituminous coating containing at least 0.048 kilogram of metallic pigment per liter of coating (0.4 pound per gallon) including, but not limited to, zinc pigment.

Table 1. Volatile Organic Compound (VOC), Content Limits for Architectural Coatings

Coating category	Grams VOC per liter	Pounds VOC per gallon
Metallic pigmented coatings	500	4.2

International Zinc Association

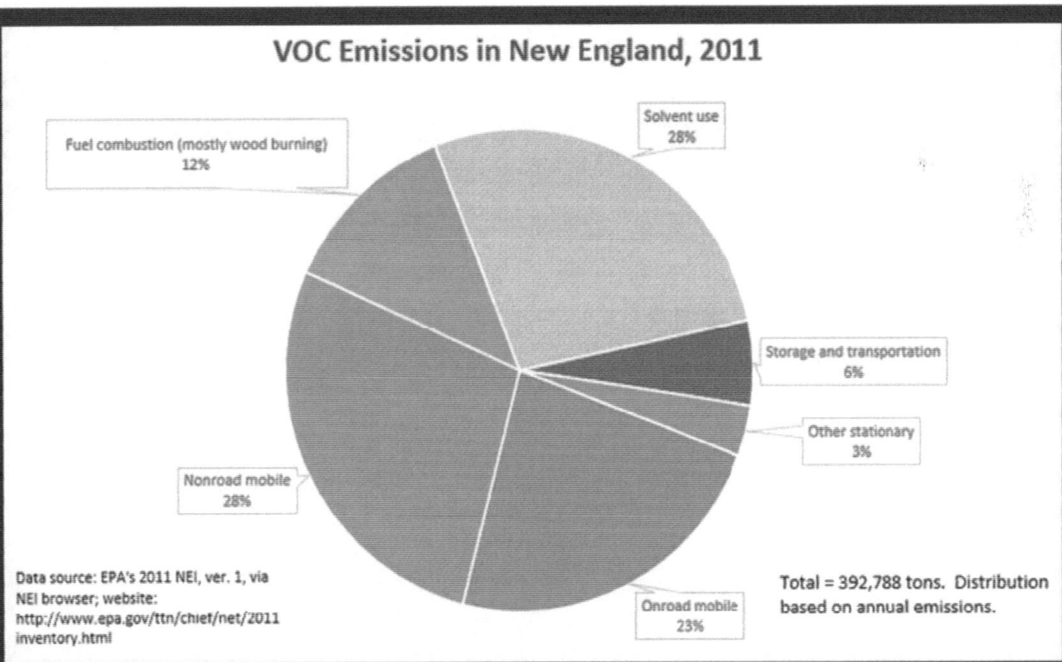

VOC's react with nitrogen oxides on hot summer days to create smog

International Zinc Association

ZC-292 Level of Zinc Required in Zinc-Rich Paints

1980-1994

Steel Structures Painting Council

- 36 paints exposed at 2 outdoor exposure sites (severe industrial and marine), also 2 accelerated tests
- Organic, inorganic, solvent reducible, water reducible, post cured, 1K and 2K formulations represented

International Zinc Association

SSPC –PS 12.00 September 1991

- Inorganic paints: minimum Zn is 74% by weight in the dry film
- Organic paints: minimum Zn is 77% by weight in the dry film
- Since then this standard has been revised "Zinc-rich coatings are successfully formulated with zinc dust levels ranging from 65 to 95 percent by weight in the dried film"

International Zinc Association

VOC's in SSPC PS 12.00 (1991)

- Class 1 $\leq$ 60 g/L (0.5 lb/gal)
- Class 2 $\leq$ 120 g/L (1.0 lb/gal)
- Class 3 $\leq$ 250 g/L (2.1lb/gal)
- Class 4 $\leq$ 340 g/L (2.8 lb/gal)
- Class 5 $\leq$ 420 g/L (3.5 lb/gal
- Class 6 Unrestricted

- Current " see state regulations, paint label, MSDS"

International Zinc Association

ZC-356 Zinc Oxide in Water-Based Paints

1987-1988

Prof. Max Kronstein

Manhattan College, New York City

Remember from chemistry class:

- Waterborne paints still contain solvents, just to a lesser amount than solvent based paints (includes inorganic Zn-rich paints)
- Anything "water-based" can be dissolved in water

International Zinc Association

Summary

- Water-based paints with ZnO have a tendency to gel when stored at ~52C (125F) for a few months
- ZnO is nearly insoluble in water at 25C, but reacts with water at 52C, releasing some reaction products that make complex formations in water-based paint

-This project treated ZnO by either ball milling with water at 52C, or in boiling water, recovering by filtration and then mixing into paint

-Ball milling was also tried with a water-based modifying vehicle from the group of polyvinyl-acetate latex, acrylic latex, or water-dispersed aliphatic or aromatic urethanes, recovering by filtration and then mixing into paint. Aliphatic urethanes worked best.

International Zinc Association

Example – preparation of modified ZnO

- 100g ZnO "Kadox" heated in 52 °C water for 15 minutes
- 25g water-dispersed aromatic urethane R940 (a colloidal dispersion in water)
- 60g water
- Recovered, rinsed and dried

International Zinc Association

Example (cont'd) – paint formulation

- 26g modified ZnO
- 52g water-dispersed aromatic urethane R940
- 4.6g water
- 1g antifoaming agent NGX

Exposed at 52°C in oven on April 11

On May 21, it still stirred up well

Further good results for 11 months

International Zinc Association

Making a water-based zinc-rich paint

- Untreated Zn powder reacts with water to make hydrogen gas
- To achieve a stable one-component water-based Zn-rich paint, this reaction must be stopped
- But, the galvanic property of Zn must be available after the paint is placed in service

International Zinc Association

Nakamura and Verbiest, JPCL August 2017

The intermediate step: a 2K water-based Zn-rich paint

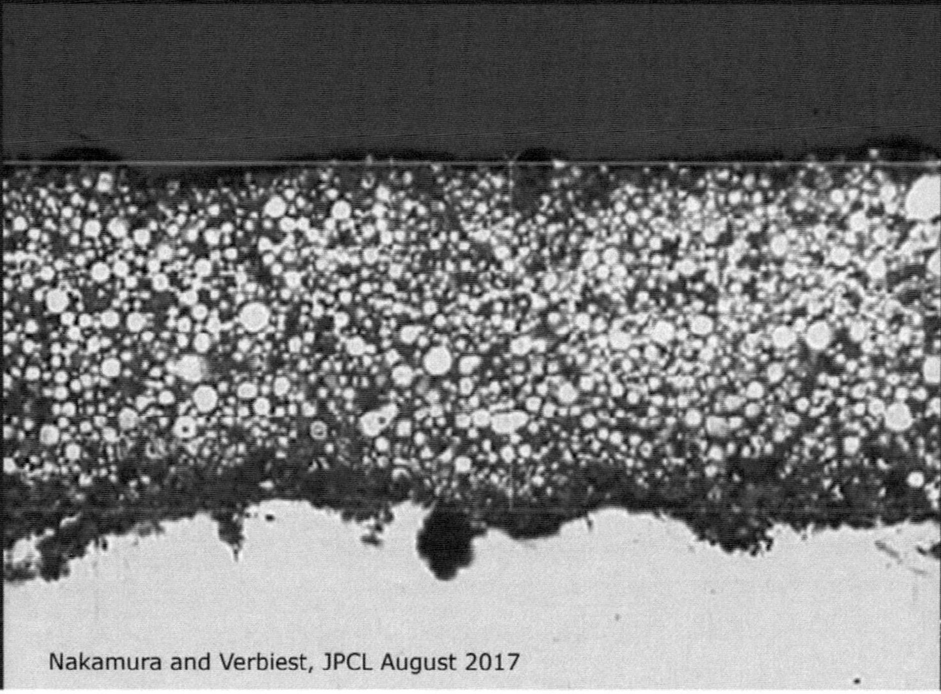

Nakamura and Verbiest, JPCL August 2017

Characteristics of 2K water-based Zn-rich paint

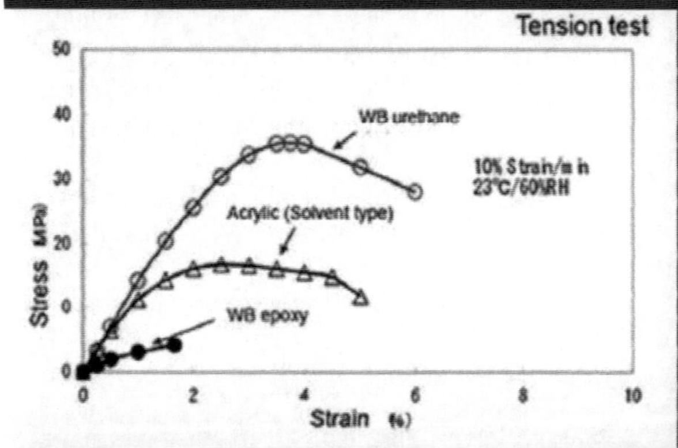

Waterbased urethane dispersion was selected as most suitable

Coating of Zn powder to improve stability

International Zinc Association

Nakamura and Verbiest, JPCL August 2017

Effect of additive level on performance (2K paint)

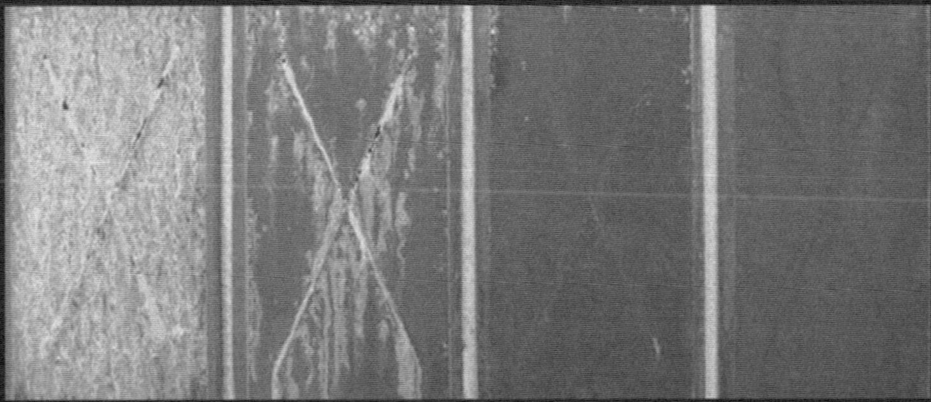

Additive results after 1,000 hours of salt-spray testing (SST). Amount of additives from left: 200 ppm, 1,000 ppm, 2,000 ppm and 3,000 ppm.

International Zinc Association

Nakamura and Verbiest, JPCL August 2017

Preventing "flash rust" on clean steel after applying WB paint: minimize water content (2K paint)

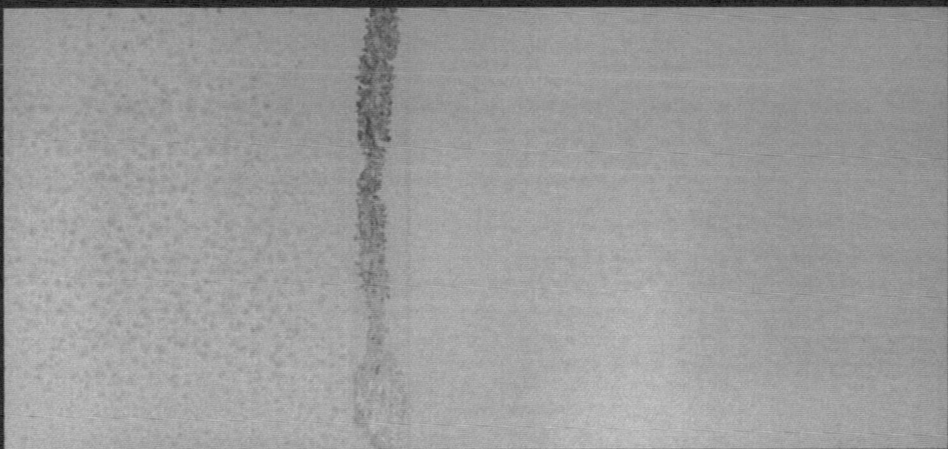

Aqueous anti-flash rust additives decreased flash rust. The left side of the image is before use; the right side, after.

International Zinc Association

Nakamura and Verbiest, JPCL August 2017

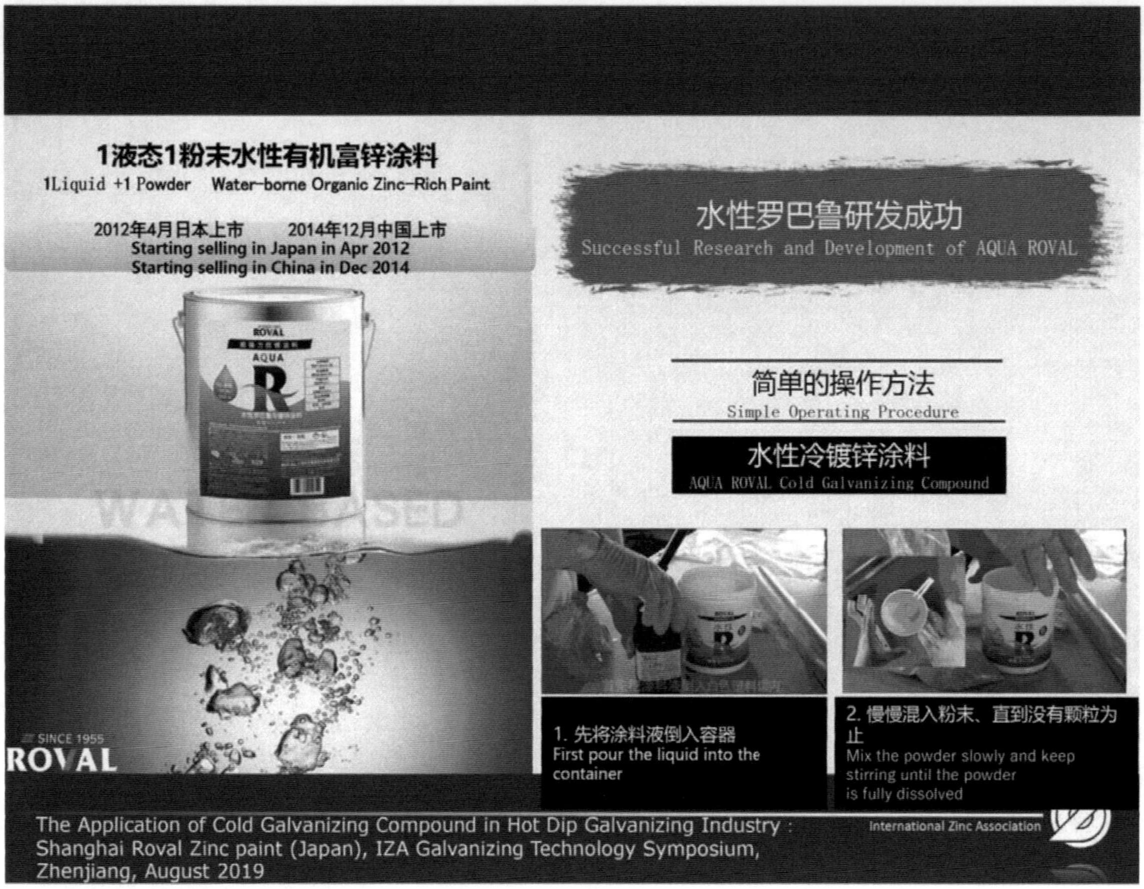

和溶剂型罗巴鲁相比

AQUA ROVAL VS ROVAL

	罗巴鲁 ROVAL	水性罗巴鲁AQUA ROVAL
安全性 Safety	△ 易燃 Inflammable	○不燃 Nonflammable
异味 Odor	△ 溶剂味 Solvent odor	◎ 基本无异味 No odor
VOC（溶剂发散量） （Solvent Divergence）	420g/L以上 >420g/L	38g/L
涂料形状 State	1液 1 liquid	1液1粉末 需要混合操作 1 liquid + 1 powder Mixed operations are required
使用期限 Valid Time	无 N/A	混合后8小时以内 8 hours after mixing

SINCE 1955
ROVAL

The Application of Cold Galvanizing Compound in Hot Dip Galvanizing Industry :
Shanghai Roval Zinc paint (Japan), IZA Galvanizing Technology Symposium,
Zhenjiang, August 2019

Characteristics of a 1K paint

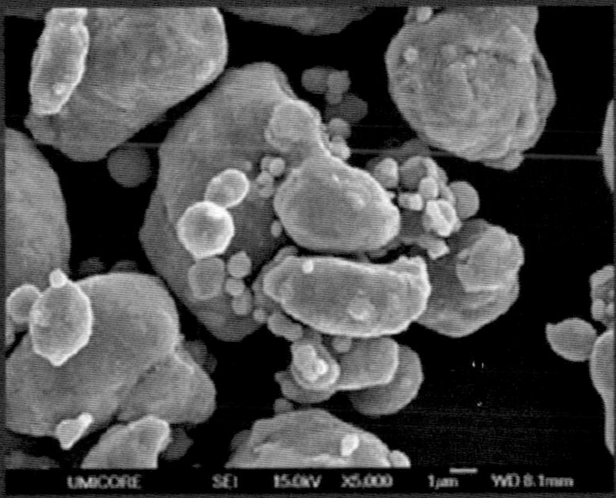

SEM image of zinc powders with the 0G treatment.

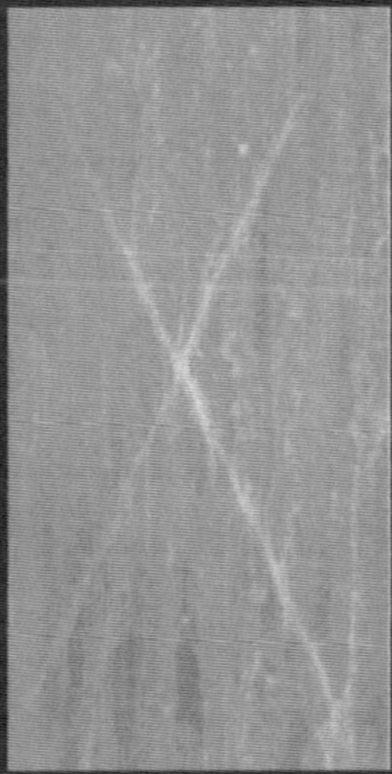

1,000 hours SST two months after mixing.

International Zinc Association

Nakamura and Verbiest, JPCL August 2017

The case for nano-zinc powders

- Highly efficient galvanic activity (at reduced film permeability) is often due to the better connectivity of the pigments because of the lower average distance between the grains
- Interparticle distance decreases with decreasing size of the particles resulting in decreasing electrical percolation thresholds as a consequence of increasing spatial density of electrical coupling connections and infinite volume spanning cluster formation of the pigment-particles.
- Osmosis-driven blistering should be absent because of the low coarseness, high packing density in the coating.
- This is obviously due to the large size of dense pigment clusters allowed by the favorable ratio of micron- and nano-size spherical grains despite the overall compositions close to the critical PVC (pigment volume concentration).

Gergely et al. Intelligent Coatings for Corrosion Control, Ch. 6, Elsevier, 2014

International Zinc Association

1. Nano zinc metal powder versus “micron” zinc powder

umicore Zinc Chemicals

Characteristics	Nano zinc powder (NOZIP®)	Commercial fine zinc powder 4P32
Average particle size (µm)	0,1 µm (100 nm)	4,4 µm
Specific surface area (m2/g)	4-6	0,45
Oil absorption (g/100g)	20,4	6
CVPC (%)	39,1	68,6
Particle shape	spherical	spheroidical
Total zinc content	> 99 %	> 99 %

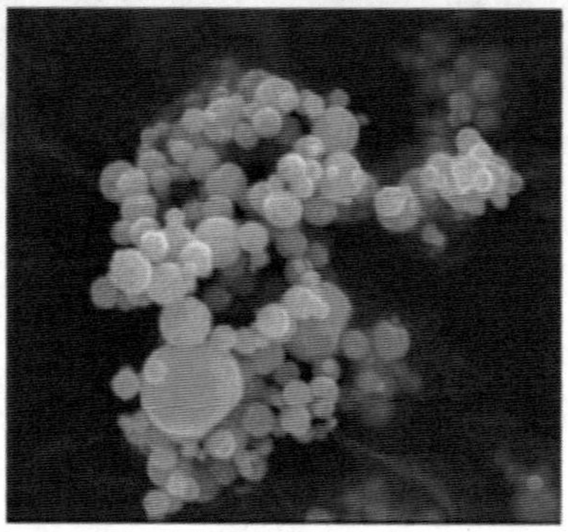

NOZIP®

Nano Zinc Metal Powder

4

International Zinc Association

Verbiest and Hochmannova, European Coatings Show, 7-9 May 2007 Nürnberg

Packing Characteristics in Powder-Air-Water System. Increasing water content traverses pendular-funicular-capillary-slurry behaviors

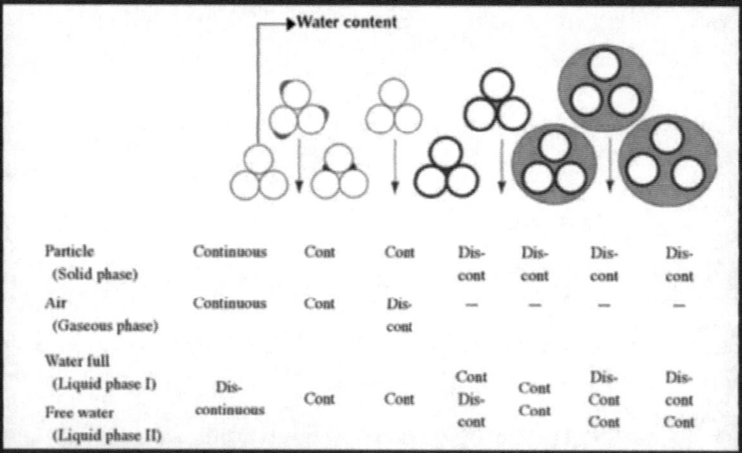

Pelletizing effect in powder-air-water system

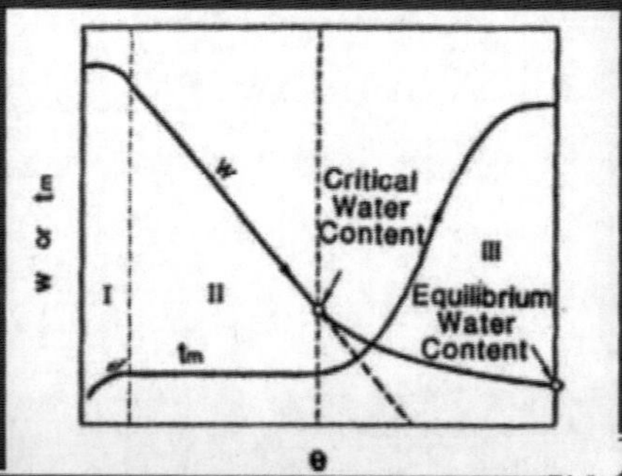

International Zinc Association

K. Umeya, J. Soc. Rheol. Jpn., **13**, 145-166, 1985

2. Paint formulation of the nano zinc primers

umicore
Zinc Chemicals

Basic concepts

❑ All zinc primers were prepared using a semisolid epoxy resin (CHS Epoxy 211) and a polyaminoamide adduct hardener (Euredur 423).

❑ Zinc primers based on the commercial zinc powder 4P32 were used as a reference

❑ Nano zinc primers were formulated based on the nano zinc metal powder NOZIP® with PVC's 1, 2 and 5

Shelf life stability of nano zinc primers

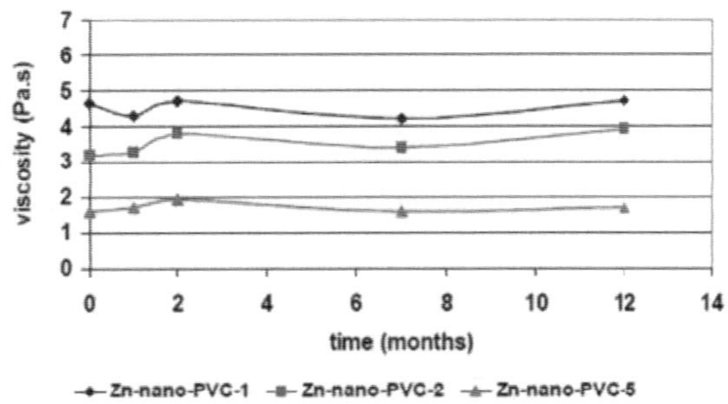

Verbiest and Hochmannova, Nuremberg, European Coatings Show, 7-9 May 2007

International Zinc Association

Zn-15%Al powder ratio in Zn-rich paint

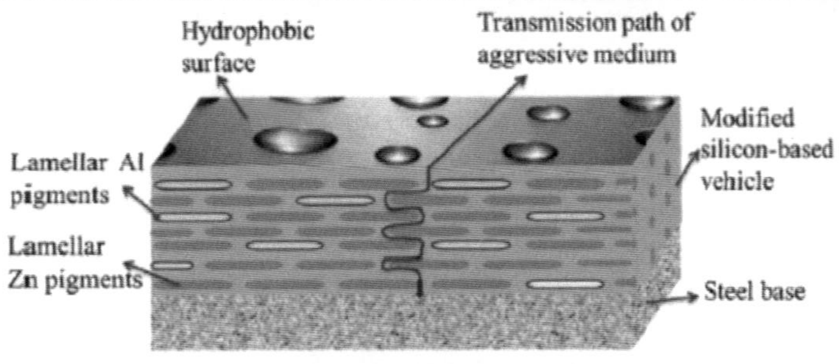

Table 3 Properties comparisons between the MFZA coating and the traditional silicate spherical Zn coating.

Properties		MFZA paint	Traditional silicate spherical Zn coating
Liquid	Pigment content (mass%)	25	85
	Suspension property	No setting	Setting
Coating	Appearance	silver	Gray
	Thickness (μm)	~60	100
	Salt spray (h)	1200	1000

Zhang et al.Progress in Natural Science: Materials International 2012;22(4):326-333

International Zinc Association

Zn-15%Al powder ratio in Zn-rich paint

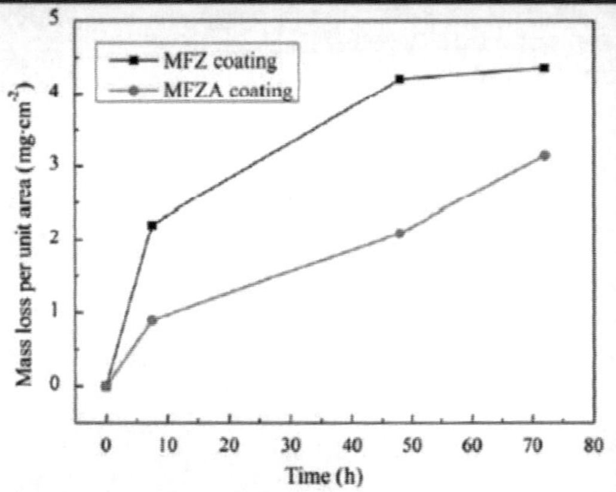

Fig. 8 Mass loss curves of the MFZ and MFZA coating samples immersed in the 0.3 mass% nitric acid solution.

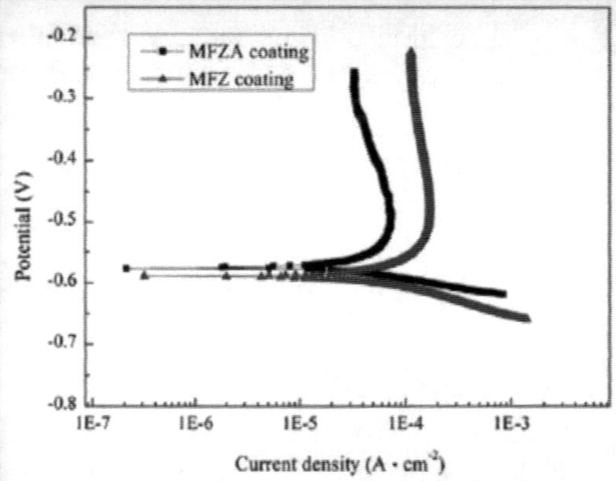

Fig. 9 Polarization curves for the MFZ and MFZA coating samples immersed in the 3.5 mass% NaCl solution for 30 min.

Zhang et al. Progress in Natural Science: Materials International 2012;22(4):326–333

International Zinc Association

Mixed Zn-ZnO compositions

Recipe of Zn–ZnO primer

Components (g)	Zn:ZnO (wt.%)								
	100:0	95:05	90:10	85:15	80:20	70:30	50:50	30:70	0:100
Zinc	90.79	86.14	81.50	76.88	72.27	63.08	44.84	26.77	00
Zinc oxide	00	4.53	9.06	13.57	18.07	27.04	44.84	62.47	88.60
Resin	5.42	5.49	5.55	5.62	5.68	5.81	6.07	6.33	6.70
Hardener	3.79	3.84	3.89	3.93	3.98	4.07	4.25	4.43	4.70
Total	100	100	100	100	100	100	100	100	100

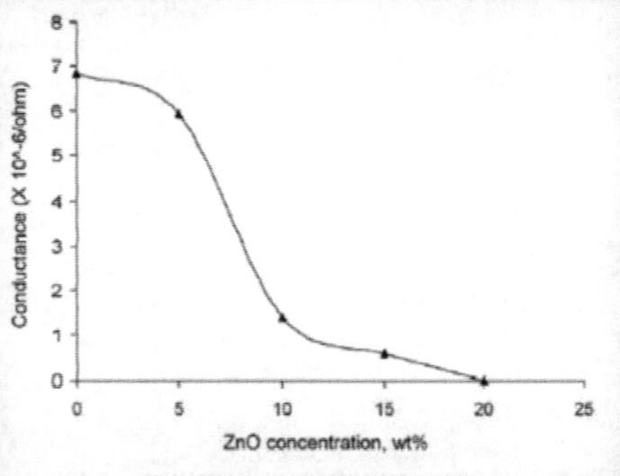

Fig. 5. Conductance vs. ZnO concentration.

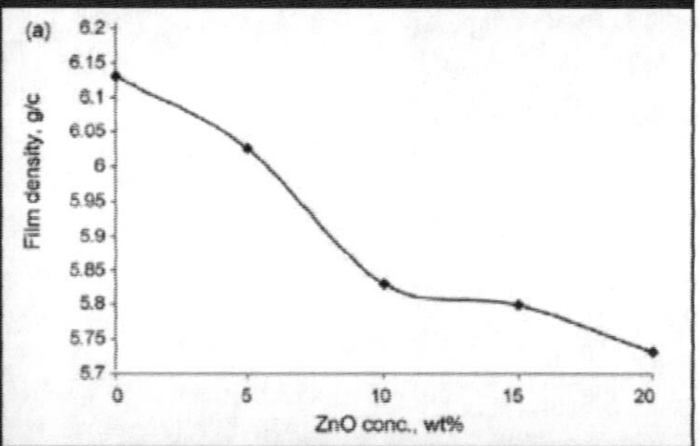

Jagtap et al. Progress in Organic Coatings 63 (2008) 389–394

International Zinc Association

Mixed Zn-ZnO compositions

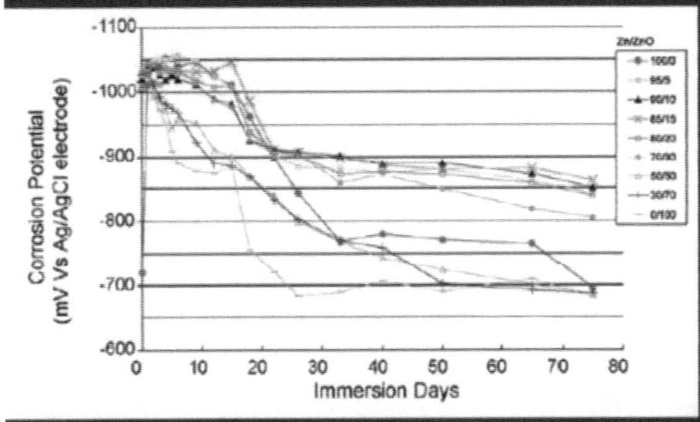

Open Circuit potential in 3.5% NaCl solution vs time

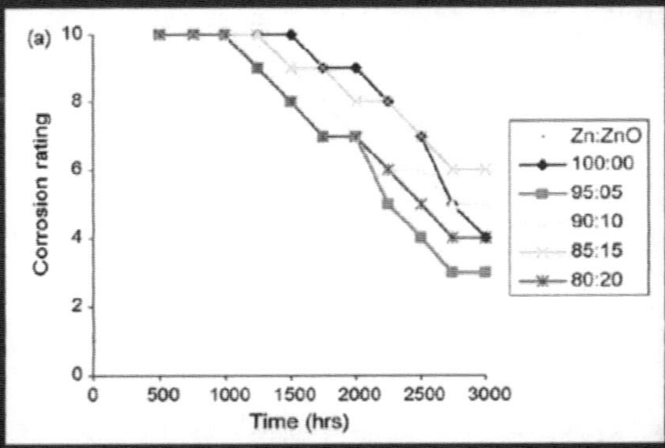

Arbitrary Corrosion Rating vs. time in 3.5%NaCl solution

The potential of Zn in sea water is approximately $-1.050V_{SCE}$, while steel has apotential of approximately $-0.650V_{SCE}$

International Zinc Association

Jagtap et al. Progress in Organic Coatings 63 (2008) 389–394

Other Extenders Carbon Nanotubes

TABLE 6.2 Composition of the Hardened Paint Coatings

Coatings	Particles	Weight% and Volume Fraction of PPy in the Primer	Weight% and Volume Fraction of PPy in the Epoxy Vehicle	Weight% and Volume Fraction of MWCNTs in the Primer	Weight% and Volume Fraction of MWCNTs in the Epoxy Vehicle	Zinc
Z	-	-	-	-	-	90 (55.8)
H1[a]	p1	0.107 (2.8×10^{-3})	0.364 (3.9×10^{-3})	-	-	70 (25.8)
H2[b]	p1	0.060 (1.9×10^{-3})	0.301 (3.2×10^{-3})	-	-	80 (37.0)
H3[a]	p2	0.144 (3.8×10^{-3})	0.493 (5.3×10^{-3})	0.450 (5.3×10^{-3})	1.542 (7.3×10^{-3})	70 (25.8)
H4[b]	p2	0.079 (2.6×10^{-3})	0.408 (4.3×10^{-3})	0.245 (3.6×10^{-3})	1.275 (6.0×10^{-3})	80 (37.0)

[a]Nano-size particles were at 3.21 and 11.0 wt.% in the primer and the epoxy binder, respectively.
[b]Nano-size particles were at 1.75 and 9.1 wt.% in the primer and the epoxy binder, respectively.

Gergely et al. Intelligent Coatings for Corrosion Control, Ch. 6, Elsevier, 2014

International Zinc Association

Other Extenders
Carbon Nanotubes

TABLE 6.1 Composition of the Nano-Size Particles in Weight % Determined by Elemental Analysis Based on the Nitrogen Element (on the Average of Three Measurements)

Samples	Alumina Content (wt.%)	Types of the MWCNTs	MWCNT Content (wt.%)	PPy Content (wt.%)	Charging Efficiency[a] of PPy (%)
p1	96.6	MWCNT-30	-	3.28	$3.7\pm5\times10^{-3}$
p2	77.9	LS NC7000	14.01	4.48	$2.1\pm4\times10^{-1}$

[a]Coulombic efficiency of PPy was derived from the voltammetry charging currents normalized to the surface of platinum working electrode and mass of measured samples (on the average of five measurements).

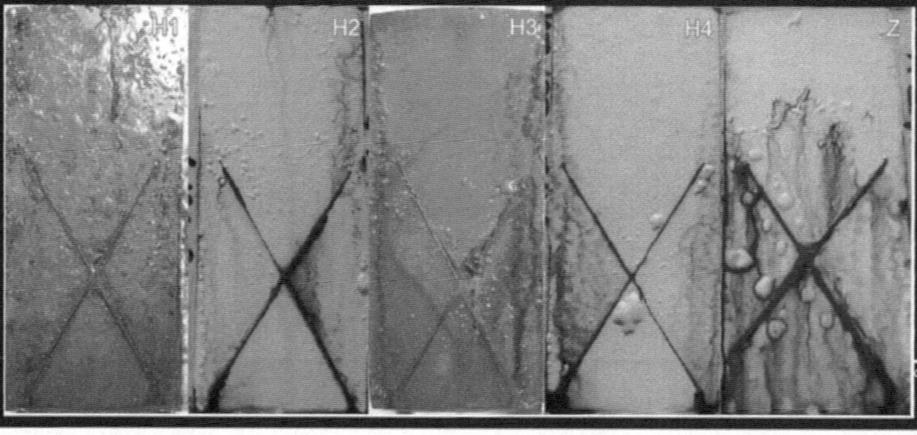

Teslan® 1102 Zinc CNT Primer

Teslan® 1102 ZN-CNT Low VOC Primer is formulated for direct application to abrasive blasted steel surfaces. It is a solvent-based two-component zinc and carbon nanotube (CNT) epoxy-polyamide primer designed to protect steel in severe offshore and coastal topside environments. Typical recommended applications include drilling rigs, offshore platforms, storage tanks, process vessels, pipelines and related oil and gas equipment.

TESLAN 1102 Zn-CNT Low VOC Epoxy Primer – Part A_REACH SDS (3-01-16)
TESLAN 1102 Zn-CNT Low VOC Epoxy Primer – Part B_REACH SDS (3-01-16)

TESLAN 1102 ZN-CNT Low VOC Epoxy Primer PDS (7-31-18)

TESLAN 1102 ISO 12944-9 ZN-CNT Low VOC Epoxy Primer PDS (07-31-18)

https://www.teslanano.com/product/teslan-1102-zinc-cnt-primer/

International Zinc Association

The solvents question

Hansen Solubility Parameters

- **The idea behind the model**

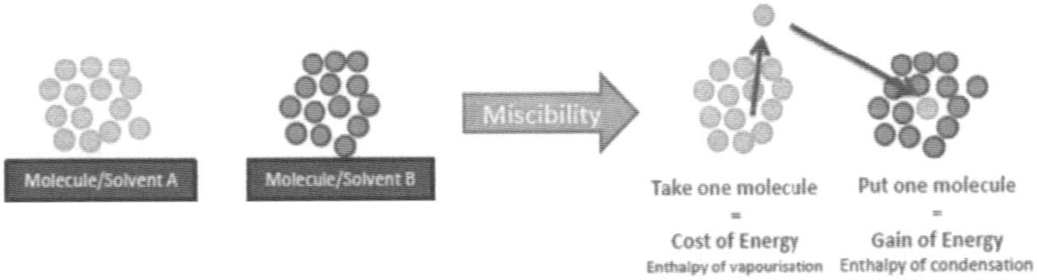

If the 2 solvents are very similar:
Enthalphy of vapourisation ~ Enthalpy of condensation → Miscible

✓ **Energy needed to take apart one molecule (into gas form) → Cohesive Energy**

- 3 components which represent the 3 forces of interaction between molecules.

$$E_c = Ed + E_p + E_h$$

Hansen Solubility Parameter

- **Cohesive energy**

E_d → London Dispersion Forces
- Temporary attractive forces due to electrons asymmetric distribution that can create dipoles

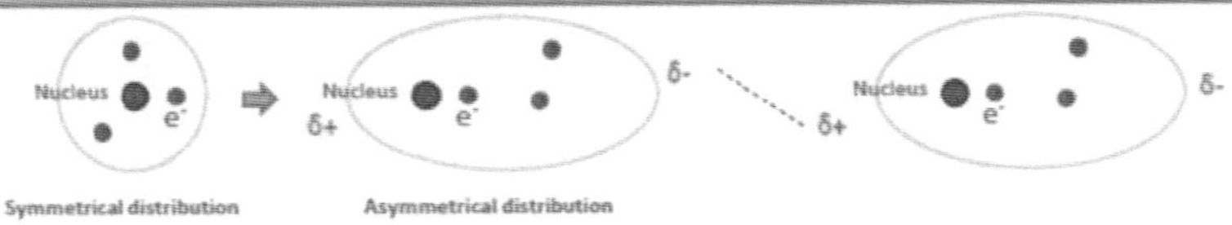

E_p → Polar Forces
- Dipole-Dipole Interactions due to unbalanced charges within a molecule

E_h → Hydrogen Bounding Forces
- Attraction between the lone pair of an electronegative atom and a hydrogen atom that is bonded to an other electronegative atom

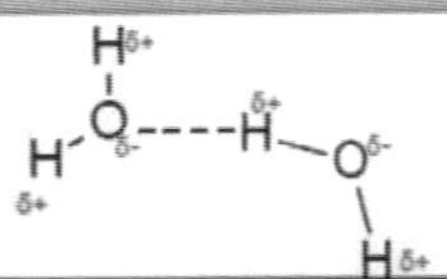

Hansen Solubility Parameters

SpecialChem
The material selection platform

- **Definition**
  - ✓ **To compare molecules of different size we can use the** Cohesive Energy Density (Cohesive Energy E_c **divided by the molar volume)**

$$\frac{E_c}{V} = \frac{E_d}{V} + \frac{E_p}{V} + \frac{E_h}{V}$$

  - ✓ **Hansen Solubility Parameters δ are defined as δ^2=E/V:**

$$\delta_t^{\,2} = \delta^2_{\,d} + \delta^2_{\,p} + \delta^2_{\,h}$$

- δ_d : Dispersion forces parameter
- δ_p : Polarity Dipole-Dipole interaction parameter
- δ_h : Hydrogene bonding parameter
- δ_t : Total Cohesive Energy Density

δ unit: $J^{1/2}/cm^{3/2}$

Hansen Solubility Parameters

SpecialChem
The material selection platform

- **Solubility volume**

Perfect match of Hansen Solubility Parameters are not required to have miscibility!

By testing many solvents with a polymer for example, a sphere can be formed containing all the « good » solvents:

Solubility Volume

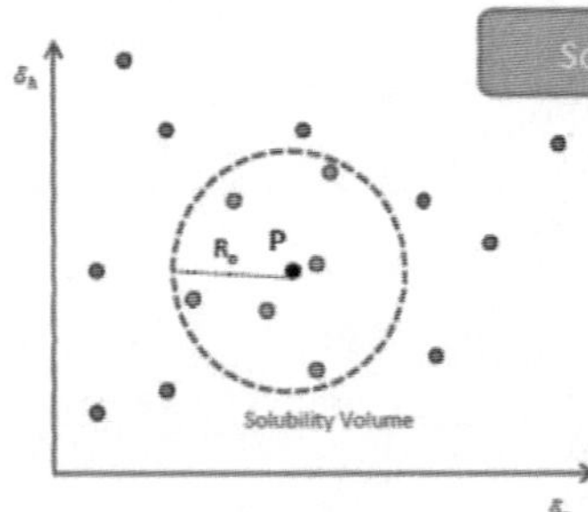

2D projection of Hansen Space

- Are "good" solvent for P
- Are "Bad" solvents for P

Definition of "good" solvent may vary:
- Swelling for reticulated polymers
- Stable suspension for nanoparticles

R_0=radius of the **Solubility sphere**
it increases with T

About Water

SpecialChem
The material selection platform

- **HSP not clearly defined**

HSP for water are not easy to measure and water behaviour depends on its environment

Material	d_D	d_P	d_H	Ro
Water — Single molecule	15.5	16.0	42.3	—
Water — >1% soluble in	15.1	20.4	16.5	18.1
Water — Total miscibility	18.1	17.1	16.9	13.0
Water — Total miscibility "1"	18.1	12.9	15.5	13.9

Hansen Solubility Parameters: A User's Handbook, Charles Hansen, CRC Press, 2007

HSP of water is still an intensive research area

Case:
Water adsorption in drying coatings

International Zinc Association

Comments

- For water-based (vs. water-borne) Zn coatings we have barely scratched the surface
- Promising directions in:
  - Zn and Zn alloy powder surface chemistry
  - Role of additives (nanotubes, ZnO...)
  - Low VOC solvent optimization
- We can expect more results in all these areas

International Zinc Association

HZL EXPANSION PLANS AND NEW DEVELOPMENTS

Harsha Shetty
Chief Marketing Officer
Hindustan Zinc Ltd

HINDUSTAN ZINC

Zinc & Silver of India

India will need Zinc

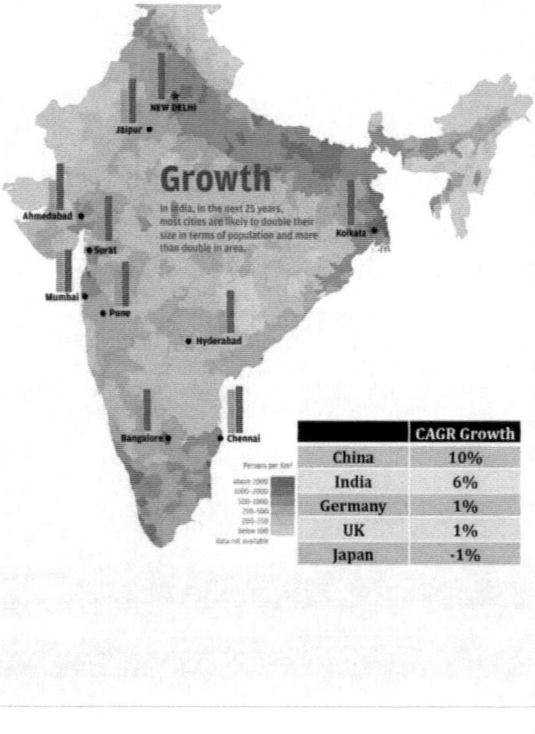

	CAGR Growth
China	10%
India	6%
Germany	1%
UK	1%
Japan	-1%

- 17% of the world's population lives in India- 1350 Million
- 5th largest economy, GDP USD 2.78 trillion, growing at 6%
- Per-capita income grew 8% in the last 5 years
- 100 smart cities under development, 400 more needed
- 2nd largest Railway network, track length of 115,000 km
- 2nd largest Steel producing nation @120 million tonne
- USD 1.4 trillion Infra spend required to reach the targeted USD 5 trillion economy size by 2025

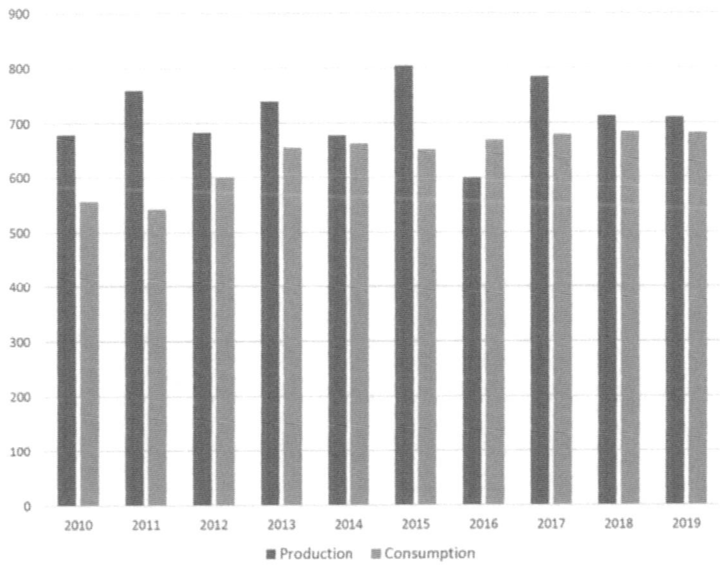

Primary Zinc in India has grew at 2.2% CAGR during CY 2010 to CY 2019 and @3.5% CAGR for Zinc in all forms

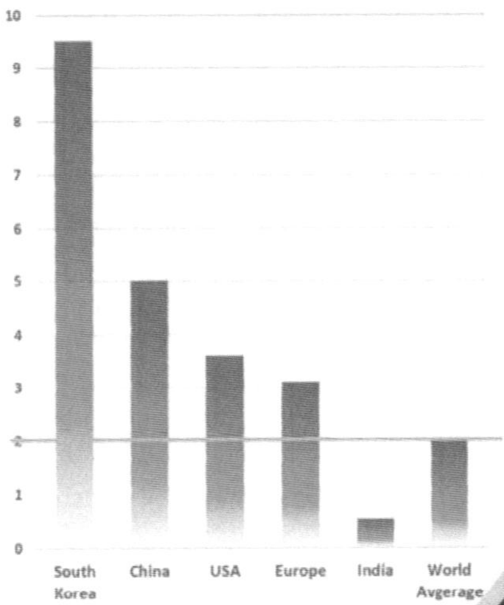

What makes India interesting to the Zinc Industry in the next decade

- Current per capita consumption : 0.52 kg
- If India can reach even half of the Global per capita Zinc consumption, India will need 1.5 MT of Zinc per annum by 2025
- The fundamental factors strongly support a spurt in Zinc requirement in the next 5 years
- HZL and IZA are spearheading the Zinc market development through focus projects to propel India to 1 Million tonne club in Zinc consumption

India's Metal Balance- Zinc in all forms.
CAGR 3.5% growth 2010-1019

		2009-10	2010-11	2012-13	2014-15	2016-17	2018-19	2019-20 T
Primary Zinc	HZL Domestic : Zinc Sales	385	412	471	483	473	513	510
	Primary Metal Prduction	**679**	**759**	**739**	**804**	**784**	**713**	**740**
	Total Primary Consumption	**522**	**500**	**574**	**617**	**655**	**647**	**640**
	HZL's Share	*74%*	*82%*	*82%*	*78%*	*72%*	*79%*	*80%*
	Alloy Market	**11**	**12**	**15**	**32**	**22**	**36**	**33**
Secondary	**Secondary Metal Production**	**43**	**50**	**49**	**34**	**34**	**35**	**35**
	Secondary (*Scrap/ Oxide/ Brass)* Imports	68	124	153	165	149	162	164
	Total Production	**722**	**809**	**788**	**838**	**818**	**748**	**775**
	Total Consumption	**631**	**666**	**780**	**858**	**875**	**905**	**887**

India has potential to be1.1 MT Zinc Market in next 4 years @5.5% CAGR growth rate

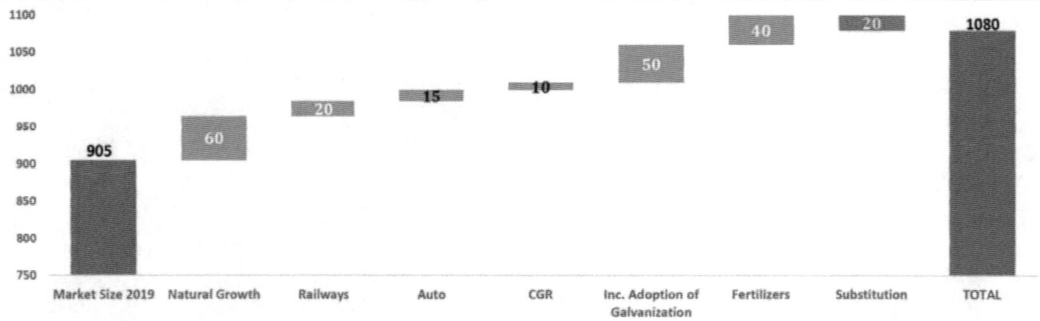

4 LEVERS TO INCREASE THE INDIAN ZINC DEMAND

RAILWAYS

- Indian Railways approves zinc metallizing of Rails
- Annual metallizing target of 3000KM, set to increase
- TOTAL RAIL NETWORK OF 1.3 LAKHS

AUTOMOBILE

- Current stand at 17% galvanization
- Target of 50% galvanization in 4 Years
- Advocacy intensified through Public media

CONT. GALVANIZED REBAR

- 1st CGR line to commence in April 2020
- CGR to reduce cost by 50%
- Opens a coated Rebar Market of 1 Million Ton

FERTILIZERS

- Government passes DBT Direct Transfer Benefit Policy in Fertilizer's
- This will give a level playing field for Zinc Based Fertilizers in India

6

Automotive
Railways
HZL- IZA market development bets
Galvanized
Rebars

Indian Auto sector to add 50KT incremental Zinc demand by 2024,
has added 50KTPA additional demand during 2012-2019
2020
2024
4 MN
Worlds FOURTH Largest
CAGR 6%
5.5 MN
24.6 MN
Worlds SECOND Largest
CAGR 7%
33 MN
1.2 MN
World's LARGEST
CAGR 8%
1.8 MN
1.1 MN
World's THIRD Largest
CAGR 12%
2 MN
AT 25% GALVANIZATION RATE
80 KT ZINC CONSUMPTION
➢ Galvanization rate has increased from 10% to 25% in the last 7 yrs
➢ Media advocacy in association with IZA intensified since Sept'19
➢ Active engagement with Ministry of road transport and Automobile companies underway
➢ Expect to increase galvanization rate from 25% to 40% by 2023
AT 70% GALVANIZATION RATE
130 KT ZINC CONSUMPTION

World's second largest Railway network set to spur Zinc demand by 20KTPA

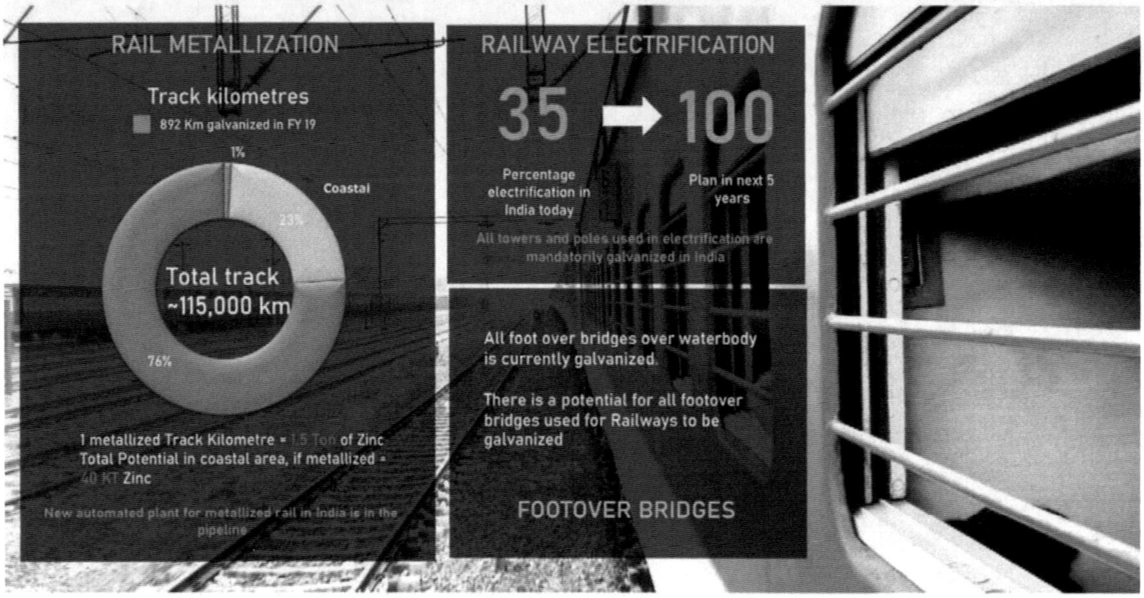

Indian Galvanized Rebar Market to take-off in 2020

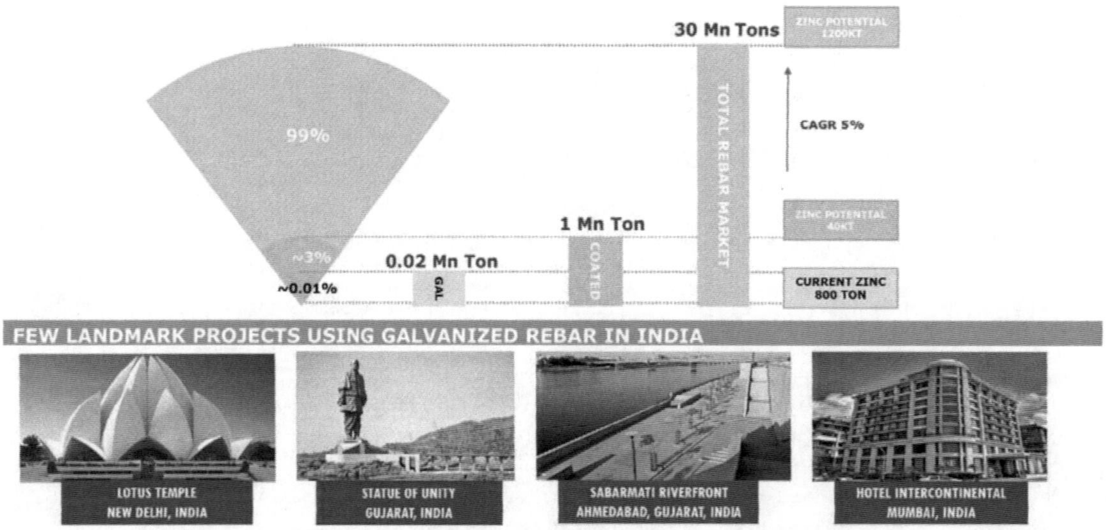

A premium Zinc-Lead and Silver company

Vedanta set to become the largest Zinc-Lead miner globally in the next 2 years

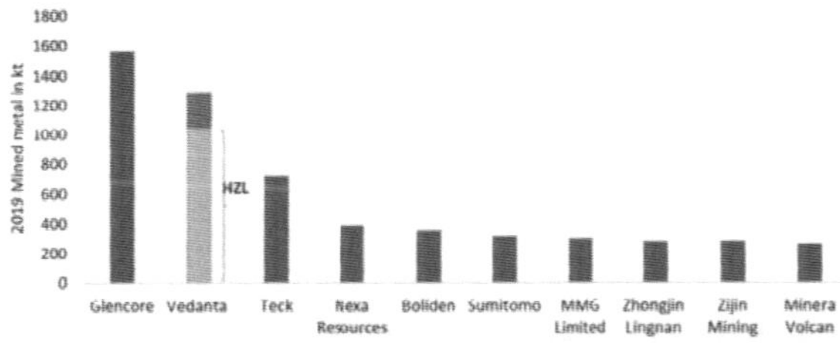

Ranked 1st in Environment dimensions in Metal & Mining sectors by *Dow Jones Sustainability Index*

Ranked 5th Globally (1st in Asia Pacific) in 'Overall Sustainability' in Metal & Mining sectors by *Dow Jones Sustainability Index*

Turnover ~ USD 3 billion, 25+ years mine life, Fastest growing silver producer in the world, 35% 5-year CAGR of UG mines

11

LEADING ZINC - LEAD MINERS
(Mined metal production in '000 MT)

LEADING ZINC - LEAD SMELTERS
(Metal production in '000 MT)

2019 ZINC COST CURVE, RANKED BY C1, COMPOSITE COST
Grouped by Company in $/t (WM2019 Q1 - All Mines)

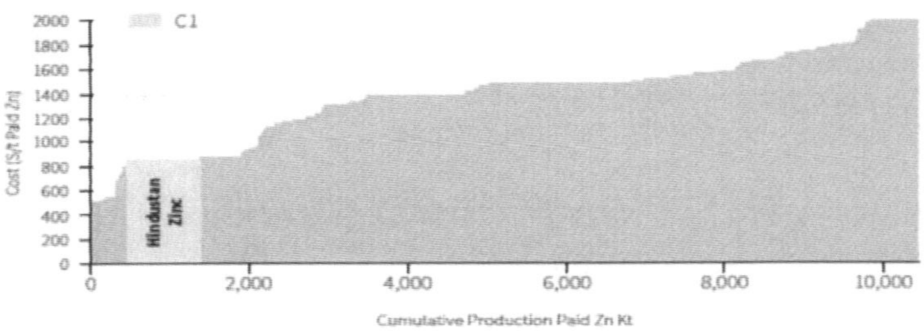

We are among the lowest cost producers of zinc globally with cost of production in (from ore to refined metals) in the first quartile of global cost curve

Enhancing the lives of communities around us

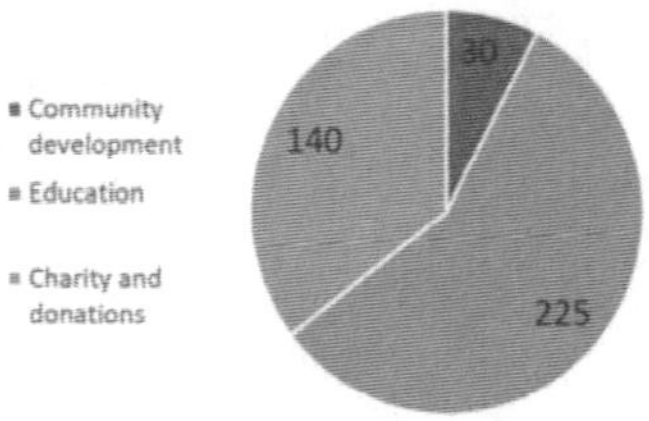

In lacs rupees

USD 0.56 million

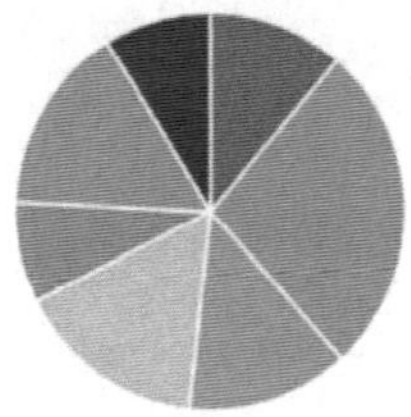

USD 18.3 million

Hindustan Zinc - World Class Assets on a growth path

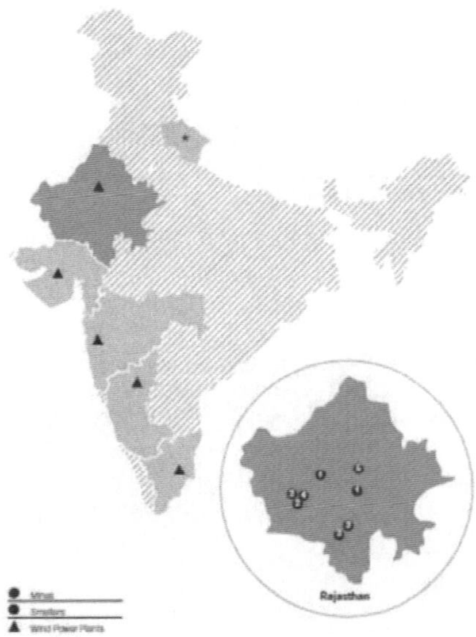

MINING

1. **RAMPURA AGUCHA**
   R&R: 90 million MT; TMC%: 15.4%
2. **SINDESHAR KHURD**
   R&R: 114 million MT; TMC%: 7.2%
3. **RAJPURA DARIBA**
   R&R: 58 million MT; TMC%: 6.8%
4. **ZAWAR**
   R&R: 105 million MT; TMC%: 5.1%
5. **KAYAD**
   R&R: 7 million MT; TMC%: 6.3%

SMELTING

1. **CHANDERIYA ZINC-LEAD SMELTER**
   Zn: 550 KT; Pb: 85 KT
2. **DARIBA SMELTING COMPLEX**
   Zn: 234 KT; Pb: 120 KT
3. **DEBARI ZINC SMELTER**
   Zn: 88 KT
4. **RENEWABLE POWER**
   Wind: 274 MW; Solar: 38 MW

R&R: 375 mMT
MINE LIFE: 20+ yrs

873,000 TPA
ZINC SMELTING

205,000 TPA
LEAD SMELTING

800 TPA
SILVER REFINING

312 MW
RENEWABLE POWER

35.4
WHRB

HZL Growth Story So far......

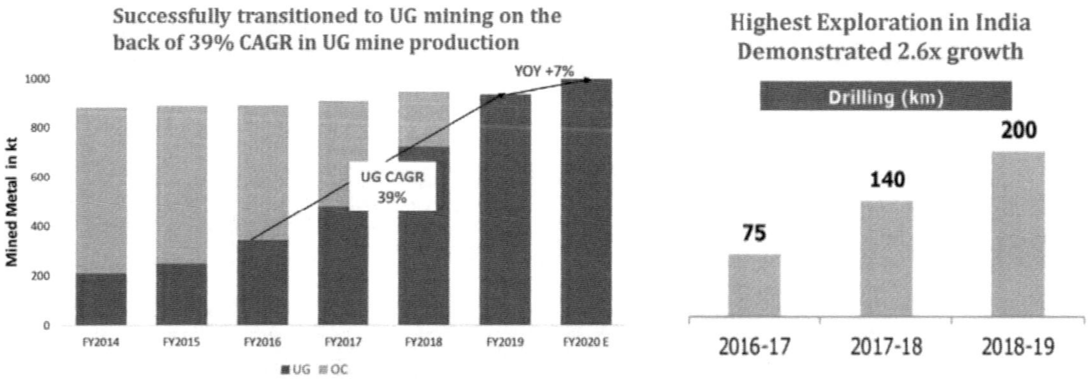

Transitioned from predominantly Open Cast to now a fully Underground mining company; Mine life of over 25 years

HZL Growth Plans: The Road Ahead...

Race to 1.5 mMT Metal, 600 kms exploration: Translated to Business Opportunities

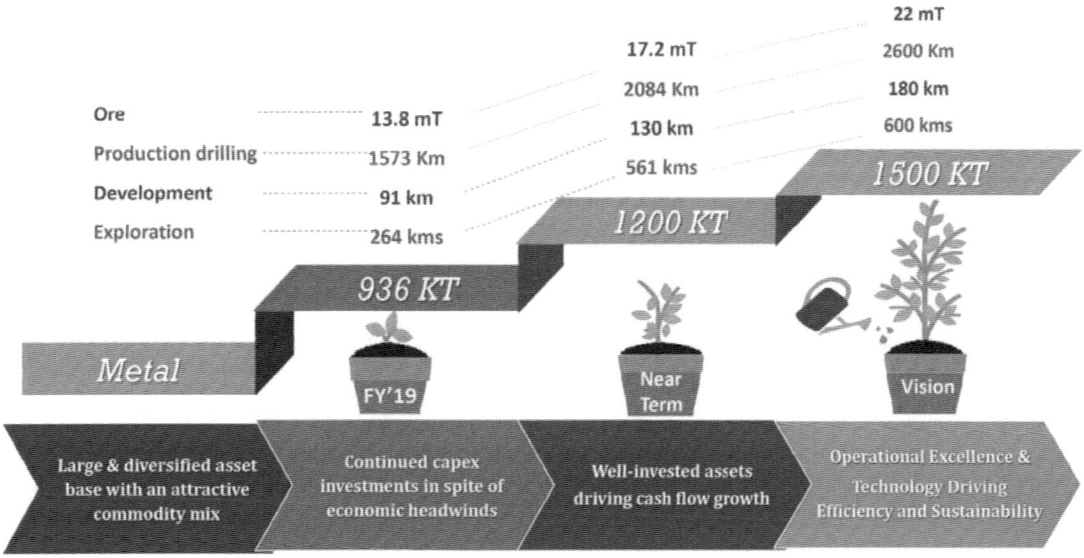

Setting Business for 1.5 mtpa and beyond

Rampura Agucha (Ore: 3.9 → 4.5 → 6)

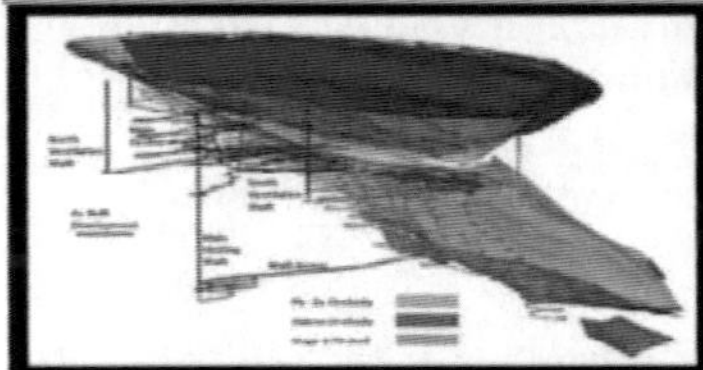

- *Shaft commissioning in H2*
- *Rapid development crew to open new levels from Q2*

Sindesar Khurd (Ore: 5.3 → 7)

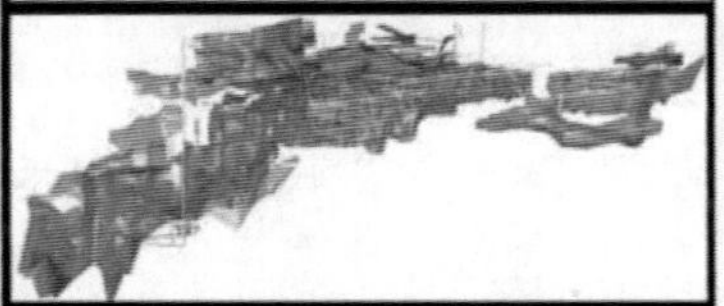

- *Largest Silver producing mine in India*
- *Shaft fully ramped-up in H2*
- *Faster paste filling*
- *Fully Digitized Mines*

Zawar (Ore: 2.8→ 3.5 → 8)

- *Commissioning of two paste fill in Q3*
- *Ballaria ramp-up in Q4*
- *Pillar mining*

Kayad (Ore: 1.2, extending mine life)

Near Mine Exploration

Rajpura Dariba (Ore: 0.9 → 2.8)

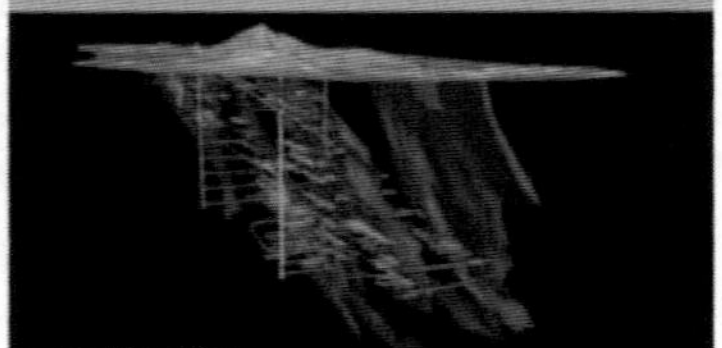

Debottlenecking

Strong foundation driving growth and transformation

R&R

600 Million Tonne

↑

400 Million Tonne

- Expanding R&R across the mines
- Best in class technology and expertise
- Largest and the most active mineral explorer in India with an exploration spend of $16 mn out of the total $50mn spent nationwide

ZINC – LEAD

1.5 MT

↑

1.2 MT

↑

1.0 MT

- Completion of 1.2 mtpa capacity in H2
- 1.5 mtpa planning underway
- Smelter debottlenecking in progress
- New resource from new ore bodies
- Focus on reserve to resource conversion

SILVER

1,000 tonne

↑

800 tonne

↑

679 tonne

- Focus on silver rich deposits
- Enhanced recoveries
- New technology - Fumer and recovery from tailings
- Natural volume growth

Inclusive growth @ HZL through Ancillaries and Zinc molten metal park

Ancillary Growth: **Chapter 1** - *Success Stories for raw materials*

1.1 Milk Of Lime

Challenge

- High Risk & unorganized players
- Frequent Quality issues & Pilferage
- Poor plant reliability (50-60%)

Strategic Engagement

Ancillary set-up backed by 10 yr offtake agreement for Milk of Lime supply in place of lime

Key Takeaways

HZL
Plant Availability: 92%
Lime Purity: 85% → 92%

Business Partner
Capital Invested: ~ 1 mn$
Est. Return: ~ 18%

1.2 Grinding of Fly Ash

Challenge

- Low blend-ability (25%) of Fly Ash with cement
- Ash grinding proven to ↑ blend-ability to 40-50%

Strategic Engagement

Fly ash grinding plant with 10 yr O&M on BOOT basis (Build Own, Operate & Transfer)

Key Takeaways

HZL
40-50% ash in place of cement
Cost Reduction: ~ $6 mn

Business Partner
Capital Invested: ~ $3 mn
Est. Return: ~ 20-25% (based on volume)

Chapter 2 - *Way ahead for Zn Metal*

2.1 Molten Zinc Metal Park

Typical Support offered by HZL in the Zinc park

- ✓ Land
- ✓ Utilities (electricity, water, air etc.)
- ✓ Long Term Contract (5yr plus)
- ✓ R&D Support

Investment brought in by Partners

- ✓ Technology, Plant & Equipment
- ✓ Manpower
- ✓ Capital Expense
- ✓ Operations & Maintenance
- ✓ All Raw Materials & MROs

THANK YOU

IHS Markit

Zinc Oxide Market Overview

International Zinc and Zinc Oxide Conferences |
Scottsdale, Arizona

February 2020

Heike Held, PhD

heike.held@ihsmarkit.com

Zinc Oxide:

Global Capacity and Production

Consumption

By region

By end use

- Rubber compounding
- Ceramics and glasses
- Agriculture
- Chemicals
- Paints and coatings
- Others: Varistors, batteries, skin care products

Global Zinc Oxide Capacity and Production

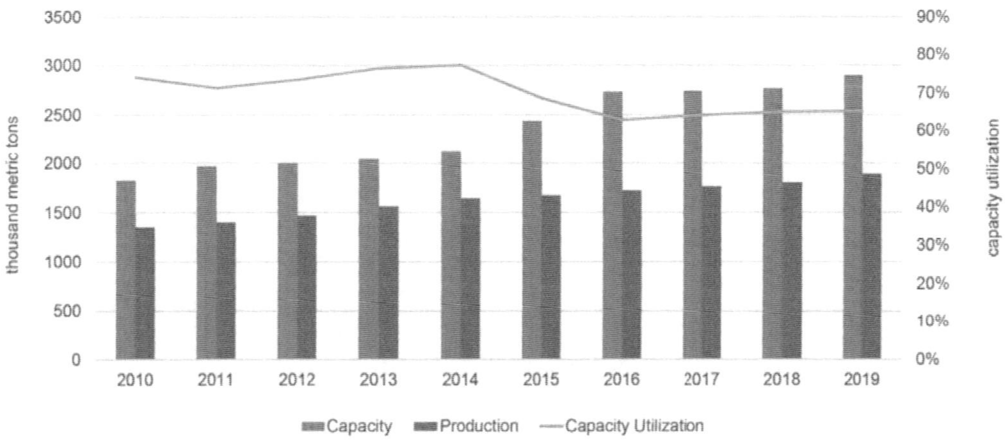

2019 Capacity Utilization by Region

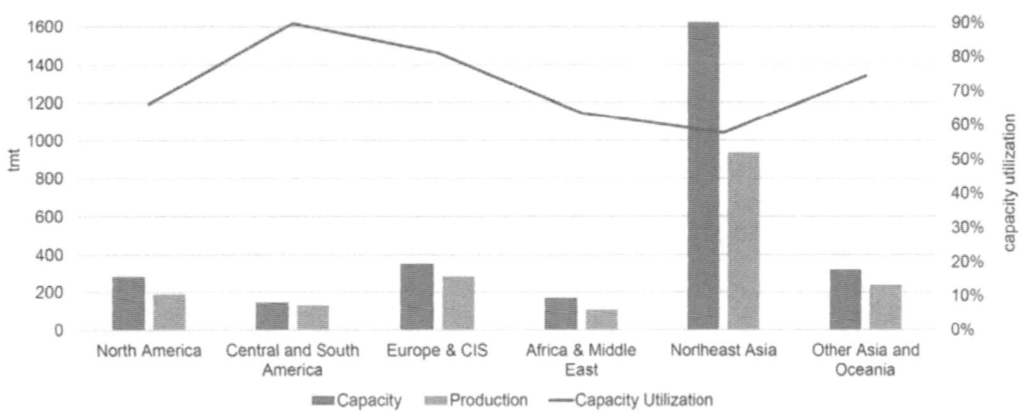

2019 Zinc Oxide Consumption by Region

Total global consumption: 1.8 million metric tons

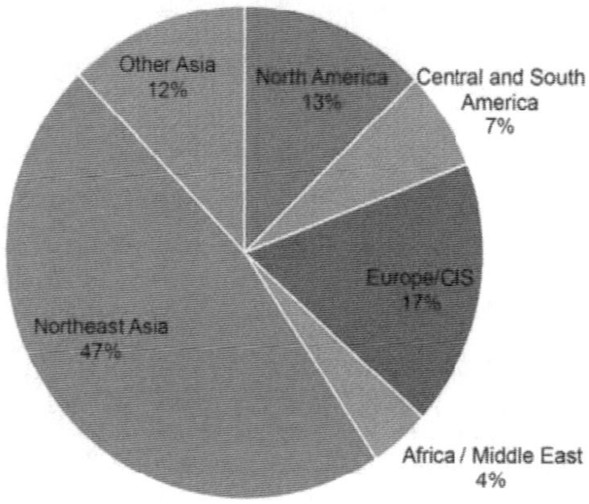

2019 Main Uses of Zinc Oxide

Total global consumption: 1.8 million metric tons

- Rubber compounding remains major end use in all regions
- Ceramics & Glasses is second largest application in all regions but North America

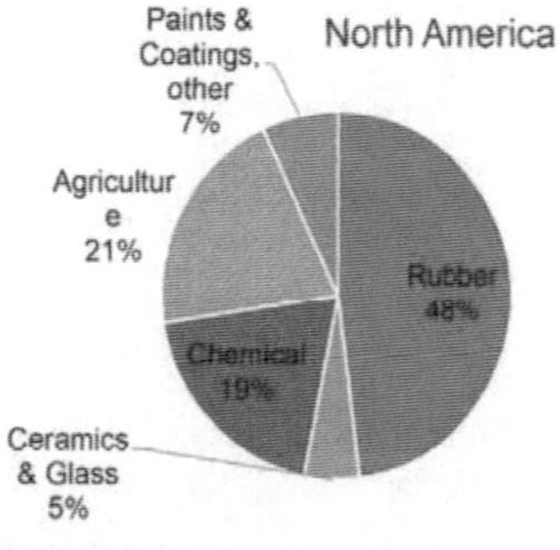

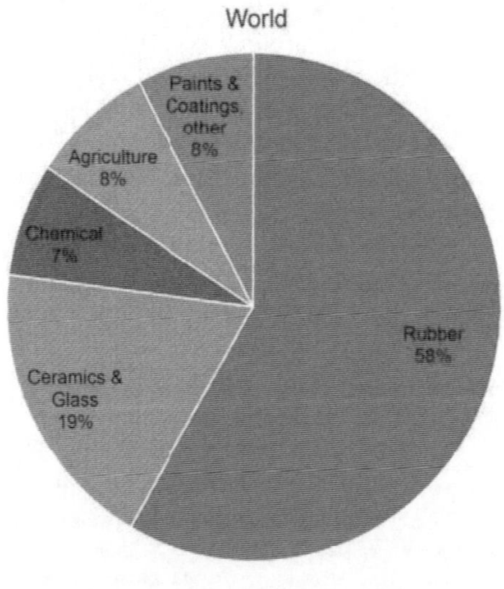

IHS Markit 7

Zinc oxide major end uses

- **Rubber compounding (58%)**
- Ceramics and glasses
- Agriculture
- Chemicals
- Paints and coatings
- Other applications

IHS Markit 8

2019 ZnO Consumption in Rubber Compounding by Region

Total consumption in the rubber segment: 1.0 million metric tons

- Approximately 75-80% of rsynthetic ubber is used in tire production.

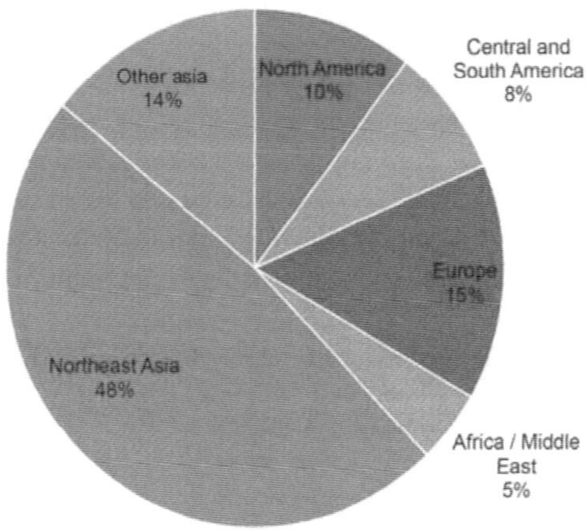

US Tire Imports and Production

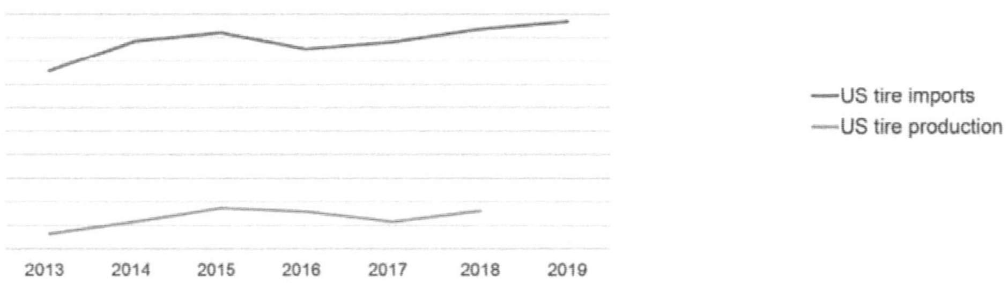

Source of US tire production: USTMA Factbook

Consumption of Polybutadiene Rubber (PBR) in Tire Production

Largest percentage growth expected in India. In Southeast Asia, the fast growth of the past five years will slow.

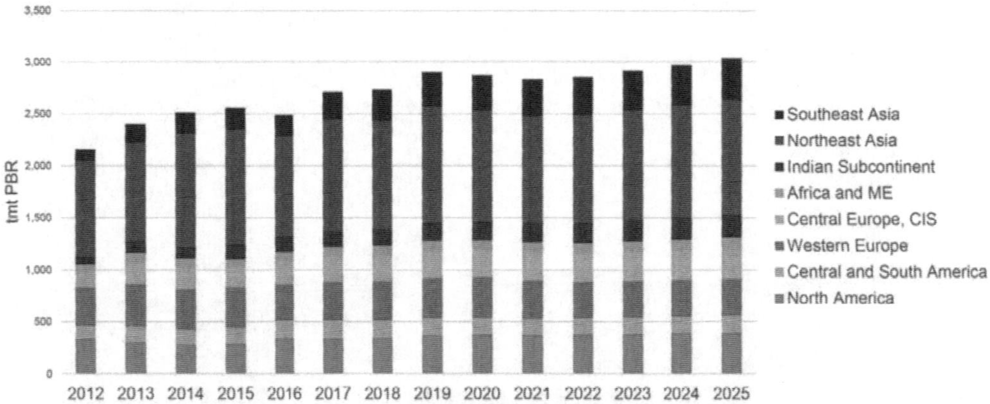

Consumption of PBR in the Production of General Rubber Goods

Largest percentage growth expected in CIS, Africa and NEA

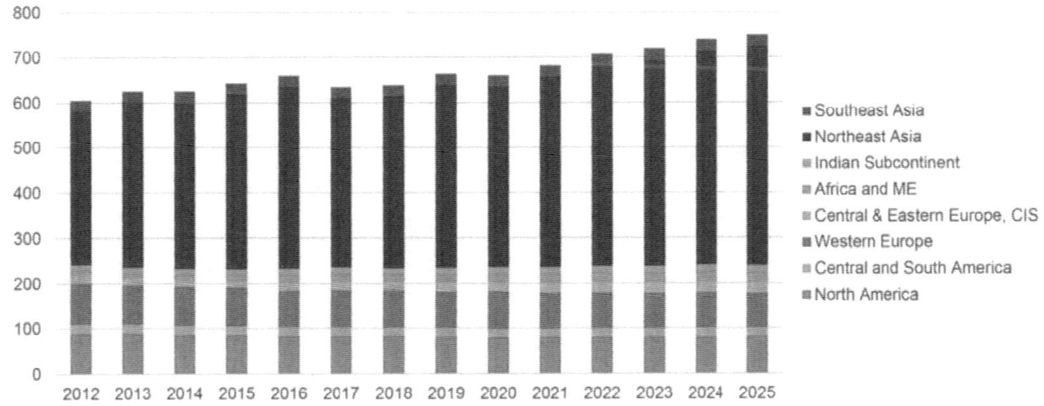

Zinc oxide major end uses

- Rubber compounding (58%)
- **Ceramics and glasses (19%)**
- Agriculture
- Chemicals
- Paints and coatings
- Other applications

2019 ZnO Consumption in Ceramic Tiles Production by Region

Total global consumption in the ceramics segment:

335 thousand metric tons

- China is still top ceramic tile producer, in spite of recent reduction of production
- Reduction in tile production in Indonesia is offset by production increase in India and Vietnam

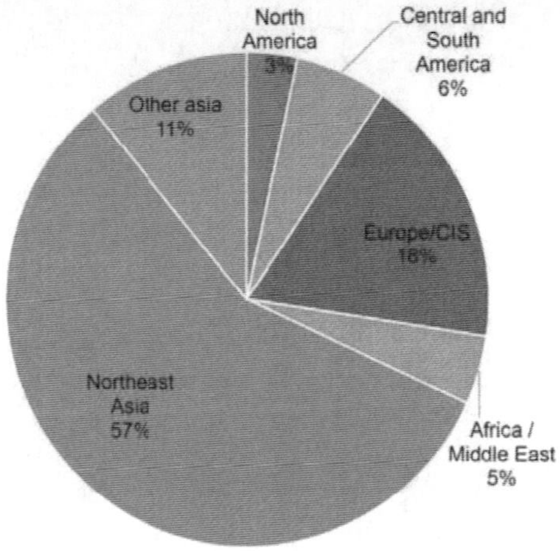

Production of Ceramic Tiles

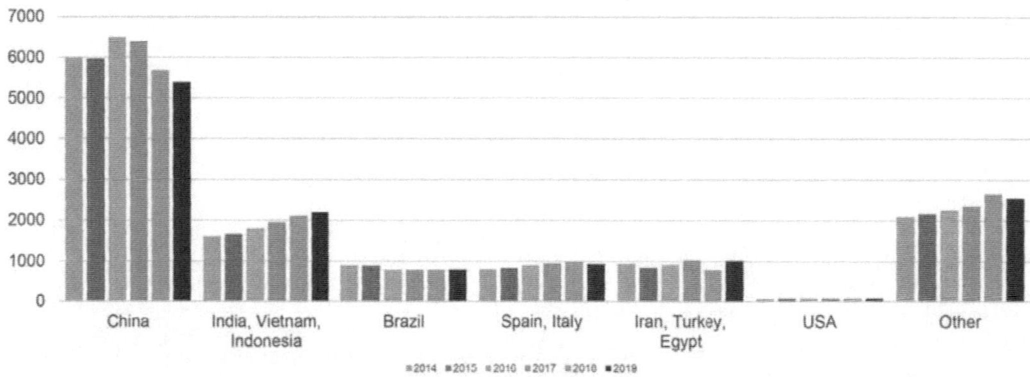

IHS Markit 15

Extra-regional Trade of Frits containing 12 – 20% ZnO

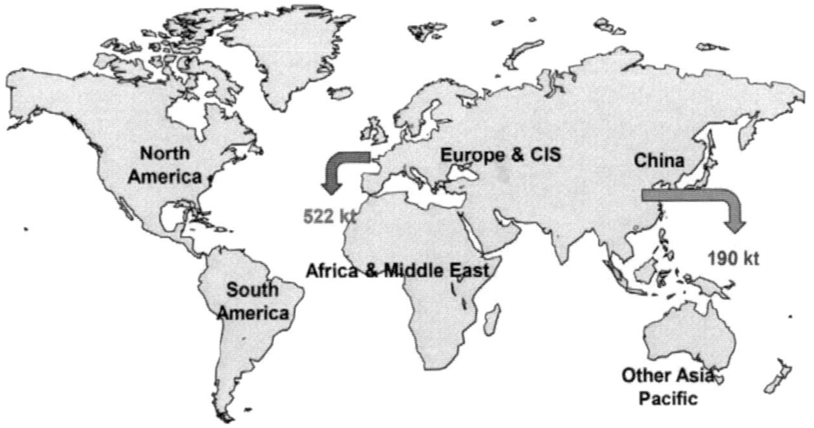

IHS Markit 16

Zinc oxide major end uses

- Rubber compounding (58%)
- Ceramics and glasses (19%)
- **Agriculture (8%)**
- Chemicals
- Paints and coatings
- Other applications

2019 Consumption of ZnO in Agriculture

Total consumption in agricultural uses: 146 tmt ZnO (117 tmt Zn)

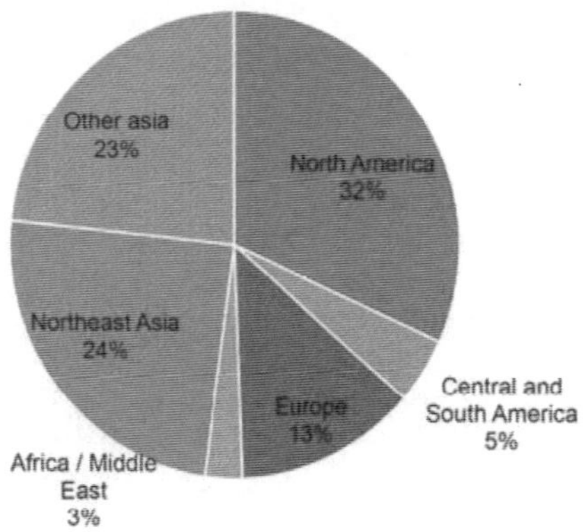

- Largest portion used in animal feed
- Substitution with zinc sulfate, especially in China
- 75% of zinc in agricultural applications is consumed in the form of zinc sulfate

Zinc oxide major end uses

- Rubber compounding (58%)
- Ceramics and glasses (19%)
- Agriculture (8%)
- **Chemicals (7%)**
- Paints and coatings
- Other applications

2019 Consumption of ZnO in Chemical Uses by Region

Total consumption in chemical uses: 129 tmt

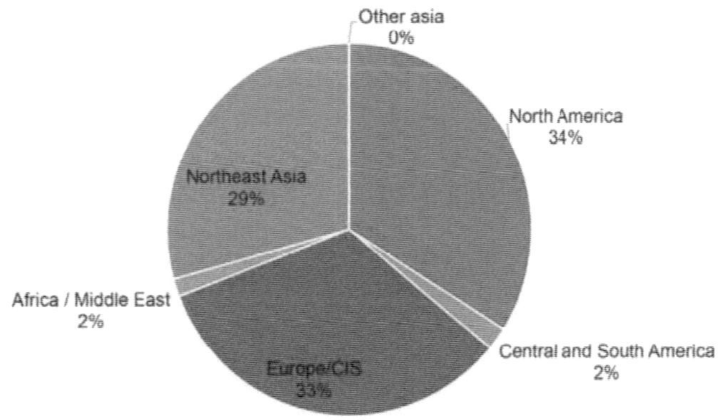

ZnO in Chemical Uses

Growth in important areas within chemical uses:

- Catalysts in desulfurization (oil drilling and refining, chemical manufacturing) → environmental protection
- preparation of
    - Zinc Phosphates (ZDDP) (lubricant additives)
      → energy conservation
    - Zinc stearate
    - Zinc octoate
    - Zinc bromide (drilling fluids)

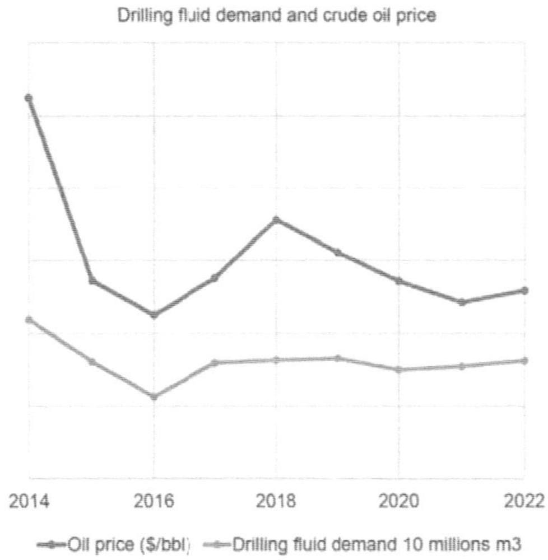

IHS Markit 21

Zinc oxide major end uses

- Rubber compounding (58%)
- Ceramics and glasses (19%)
- Agriculture (8%)
- Chemicals (7%)
- **Paints and coatings**
- **Other applications**

IHS Markit 22

ZnO in Paints and Coatings

Reduced deamnd for specialty areas of use, i.e.

- Antifouling
- Antimicrobial
- UV-protection
- Mechanical Strength of the Coating

IHS Markit 23

Zinc Oxide in Varistors

- Growing demand.
- Most of this being consumed in China.
- Strong growth expected to continue.
- Growth Drivers:
  - Continuing growth of handsets.
  - Automotive electronics growth
  - Growth of arresters with growth of smart grid technology
  - Renewable energy systems grow and require redundant circuit protection

IHS Markit 24

Conclusion:

Demand for ZnO is expected to continue to show nslow growth over the next five years.

Key demand drivers for ZnO:

- General economy and automotive market in Asia
- Recovering construction industry, tied to global increase in tile manufacturing, in spite of receding production in China
- Growing importance of zinc as micronutrient; agricultural programs especially in Asia, Africa and South America.
- Increasing demand for electronics.

THANK YOU!

CRU

Zinc Market Outlook

Zinc unlikely to prosper in the year of the rat

IZA Zinc Conference, Arizona,
16-19 February 2020

Helen O'Cleary

CRU

- Key features of the Zn market in 2019
- Mine supply
- Smelter output
- Demand
- Market balance and prices

2

Key features of zinc supply in 2019

Year-on-year change in Zn mine production

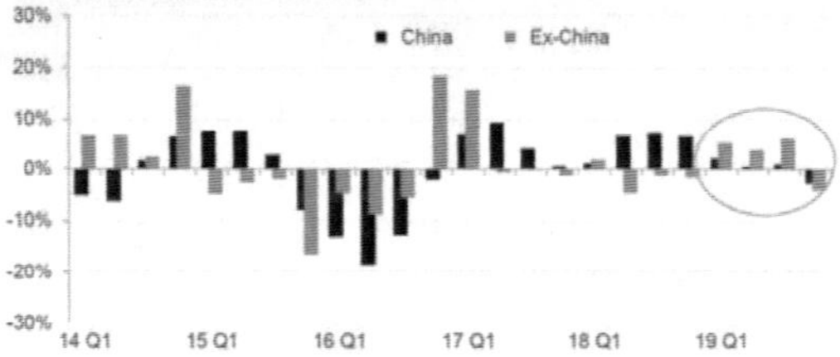

- Chinese mine supply grew by 2.5% y/y in 2019.
- Ex-China mine supply underperformed expectations:
  - HZL downgraded due to delays to Rampura Agucha underground ramp-up.
  - Gamsberg forecast reduced.
  - New Century downgraded due to ramp-up issues.

Year-on-year change in Zn smelter production

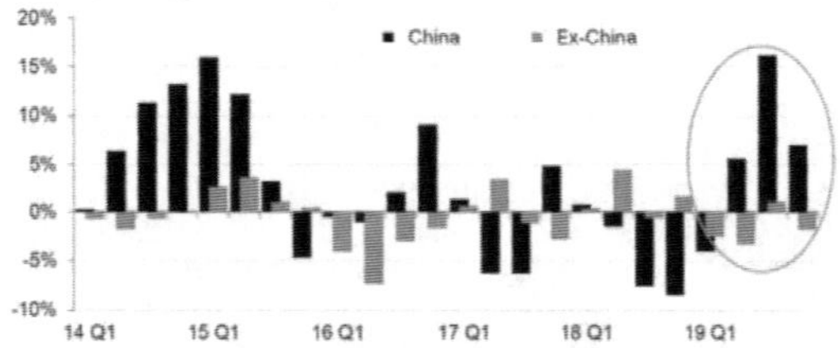

- Chinese smelters increased utilisation rates as expected after a very weak 2018.
- Ex-China smelter production disappointed:
  - Delayed restart at AZR's Mooresboro.
  - HZL downgraded due to slow mine ramp-up.

Key features of zinc demand and balance in 2019

Year-on-year change in refined Zn refined demand

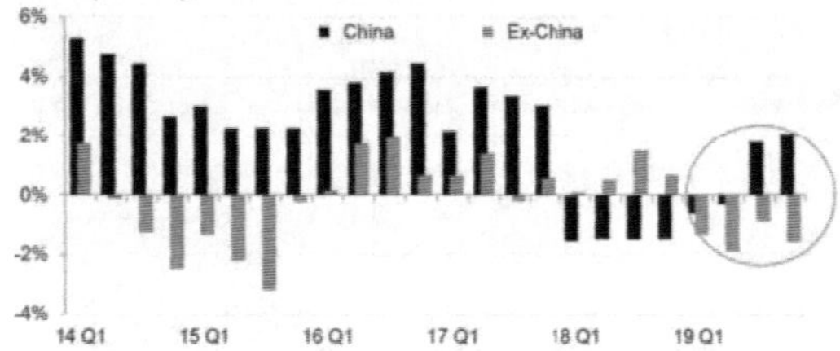

- After a weak start to the year, Chinese refined demand returned to growth in 2019 H2.
- Ex-China refined demand contracted in every quarter, due mainly to weakness in Europe.

Supply-demand balances, '000 tonnes

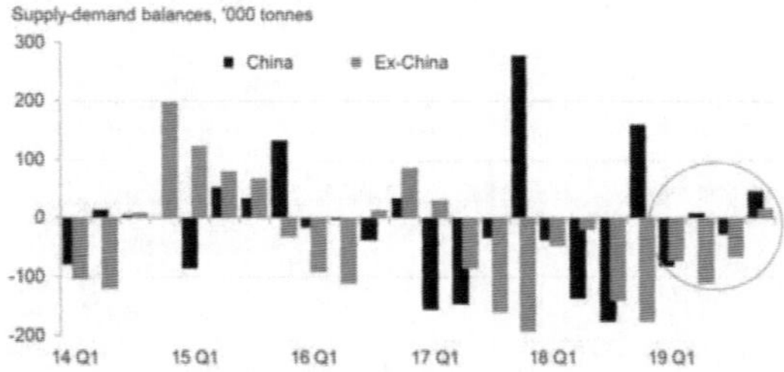

- The ex-China market registered deficits in 2019 Q1-3 and returned to balance in Q4.
- The Chinese market registered a modest surplus in 2019 Q4.

Zinc Market Outlook, January 2020

Key features of zinc stocks and prices in 2019

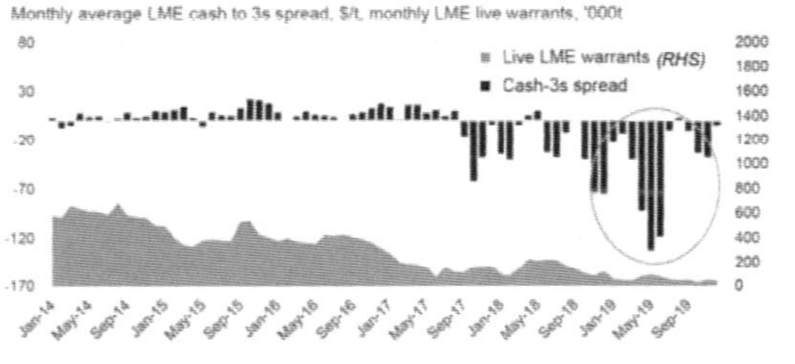

- LME stocks ended 2019 at 51,200t. Live warrants were just 42,525t.
- Record low LME stock levels led to steep backwardations in April-June 2019.

- Zinc's price trended lower for most of 2019, with US-China trade tensions playing an important role in price movements.
- Despite record low LME stocks, zincs price slumped in Q4 on higher Chinese smelter output.

5

- Key features of the Zn market in 2019
- **Mine supply**
- Smelter output
- Demand
- Market balance and prices

6

Zinc Market Outlook, January 2020

A large proportion of mine supply growth is expected in 2020

Mine supply, million tonnes zinc

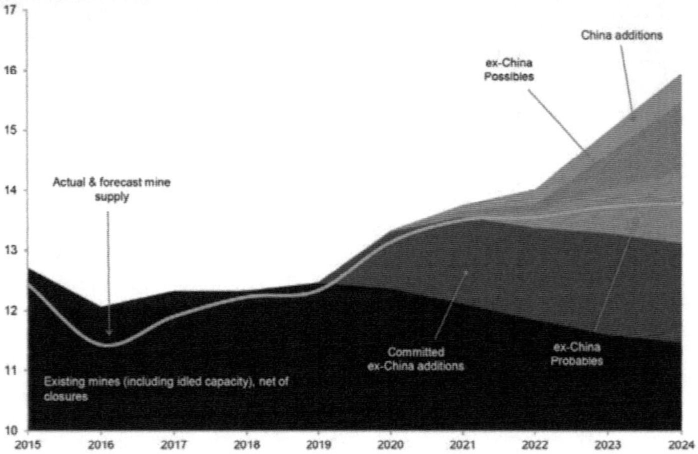

- We are forecasting an average increase in mine supply of 2.2%, to reach 13.78 Mt by 2024.
- A large proportion of this growth is expected this year.
- All gains will come from ex-China as Chinese mine supply is expected to remain flat.

Zinc Market Outlook, January 2020

Committed ex-China mine supply could add 675kt by 2024

Committed mine supply - new mines and redevelopments
Peak production 2019-2024; '000 tonnes zinc

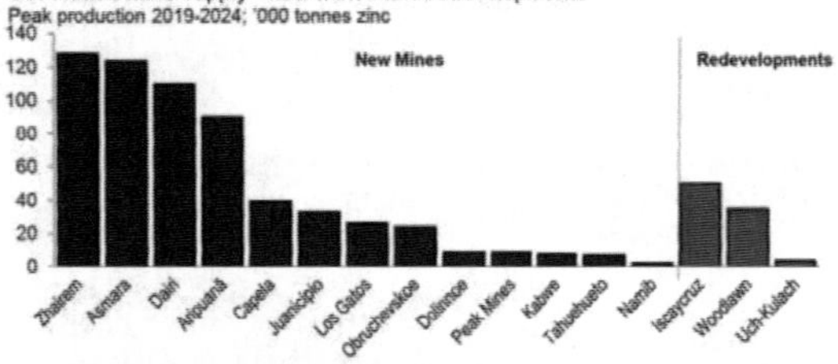

- New mines and redevelopments could add 0.6 Mt by 2024.
- Expansions and supply increases could add 1.1 Mt.
- Contractions and closures could remove 0.9 Mt.

Committed mine supply - expansions and supply increases
2019-24 production increase; '000 tonnes zinc

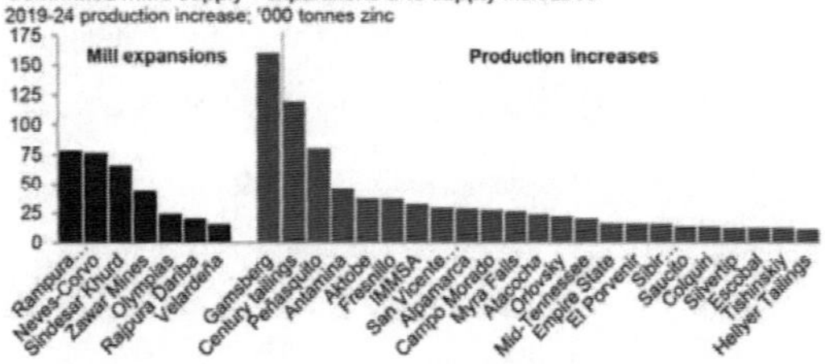

Committed mine supply - closures & contractions
2019-24 production loss; '000 tonnes zinc

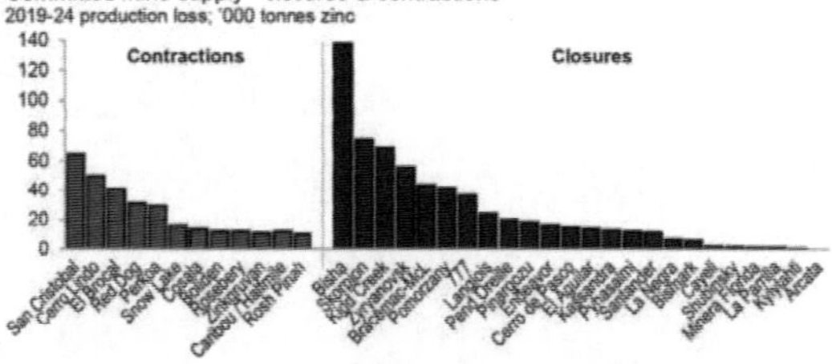

Some miners are at risk given our current price forecast

Swing producers at risk at current price forecast
LME cash (real 2019$/t Zn)

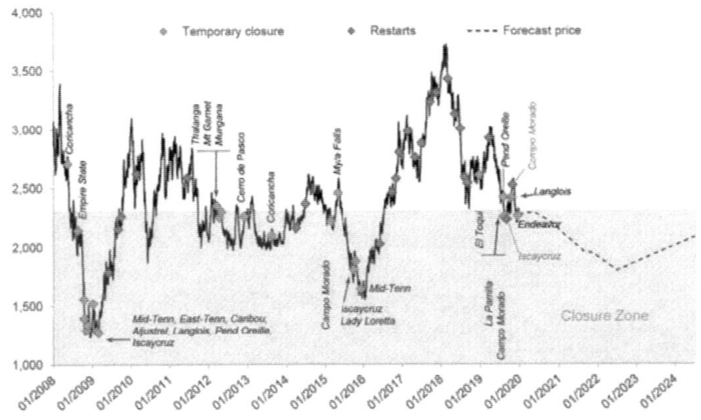

- Some swing producers have already exited.
- A number of other mines are at risk given our price forecast.
- The latest to exit, Endeavour, is not a swing producer.
- Glencore's Iscaycruz bucked the trend and restarted in September 2019.
- Telson reversed their closure decision on Campo Morado.

Chinese mine output is expected to remain flat in 2020

2020 cash cost curves for operating and firm developments
Cash costs, $/t Zn

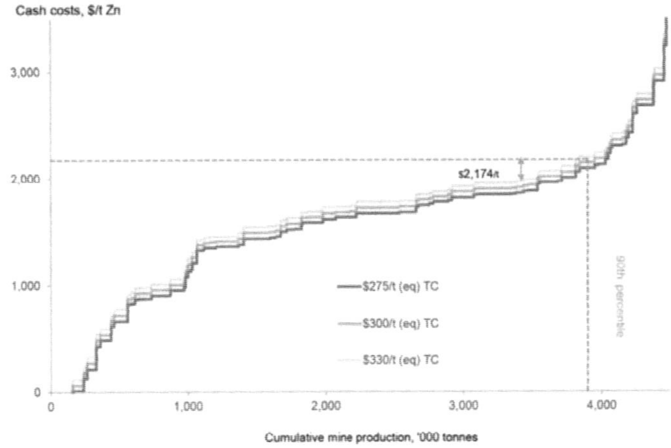

- 90th percentile of the China cash cost curve is at $2,174/t.
- Theoretical break-even point for Chinese industry.
- Some smaller mines have much higher costs than this.

Chinese mine supply could reach 4.68 Mt by 2024

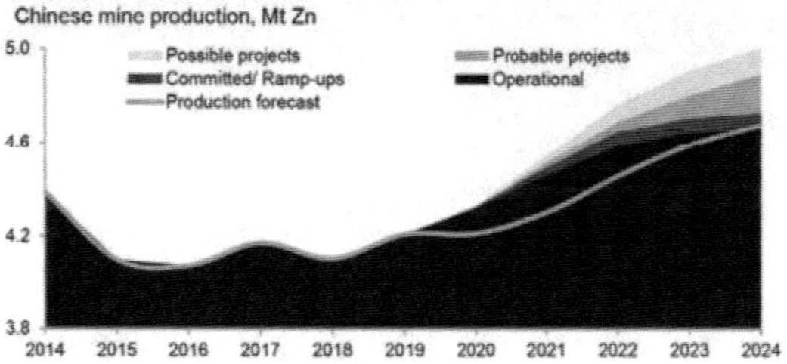

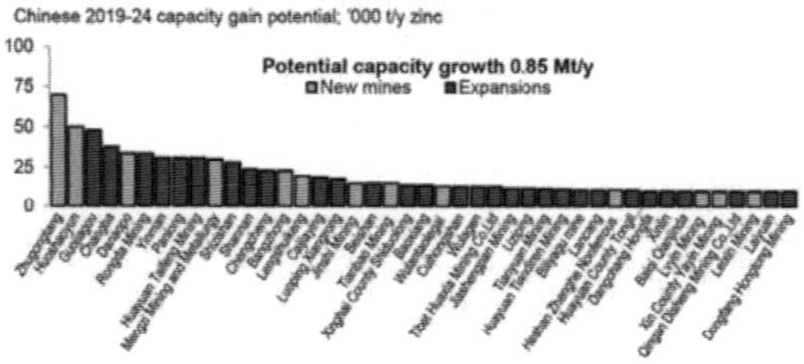

- We expect Chinese mine production to grow by 2.1% per year in 2019-2024, to reach 4.68 Mt.
- Operating mines could add 500kt by 2024.
- Currently committed greenfield projects will contribute 65kt by 2024.
- 16 Probable and Possible projects could add 274kt (pre-disruption).
- Growth over the medium term is likely to be dominated by large capacity gains.

- Key features of the Zn market in 2019
- Mine supply
- Smelter output
- Demand
- Market balance and prices

Zinc Market Outlook, January 2020

Global smelter capacity will reach 16.37 Mt in 2024

Global Zn smelting capacity, M t/y

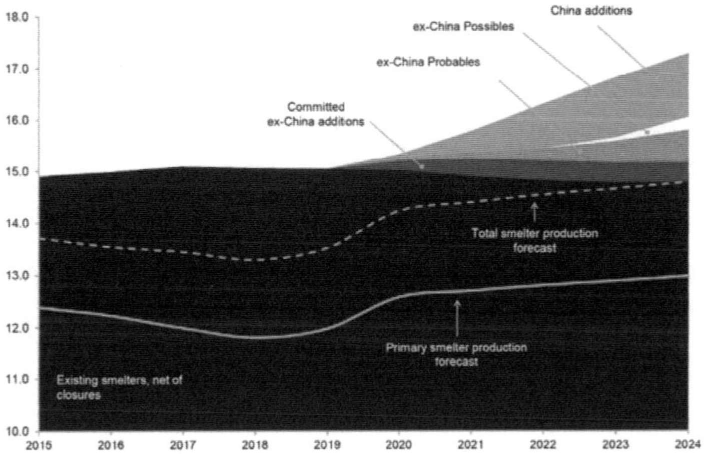

- We expect global smelter capacity growth to average 1.7% per year in 2019-2024.
- The main committed increases will come from:
  - AZR's Mooresboro
  - HZL's Indian operations
  - KZ's Townsville
- Losses include the closure of HudBay's Flin Flon in 2022.

13

Zinc Market Outlook, January 2020

We expect Chinese smelter output to continue to grow in 2020

2019 Chinese Zn smelter output, 000t

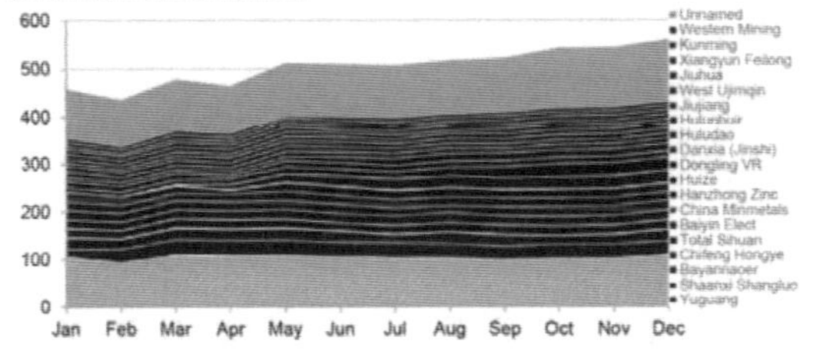

- Chinese smelter output grew as expected in 2019.
- Better environmental compliance and improved profitability helped lift output by 6.1% y/y.
- We expect 2019's 350,000t increase to be followed by a 300,000t increase in 2020 although there are downside risks to this forecast due to the novel coronavirus outbreak.

2019-24 rise in Chinese nameplate Zn smelter capacity; '000 t/y

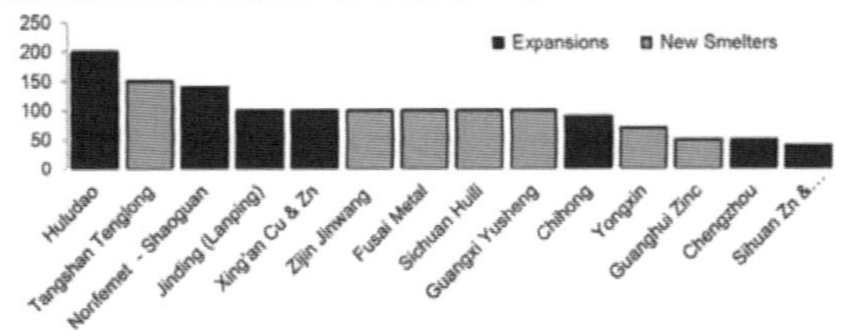

- There is a healthy pipeline of expansions and new smelters going forward.

14

Key features of the Zn market in 2019

Mine supply

Smelter output

Demand

Market balance and prices

Global demand may grow by less than 1% in 2019-2024

Refined Zn demand, million tonnes

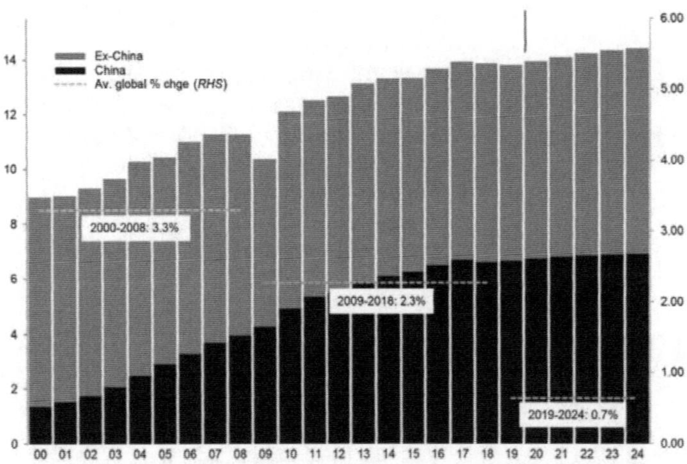

- Global refined demand growth has slowed markedly in recent years and is expected to slow further.
- We expect Chinese demand to follow a path of modest growth going forward.
- Global demand is expected to average 0.7% over our forecast.

Zinc Market Outlook, January 2020

Global galvanised sheet output is expected to grow modestly

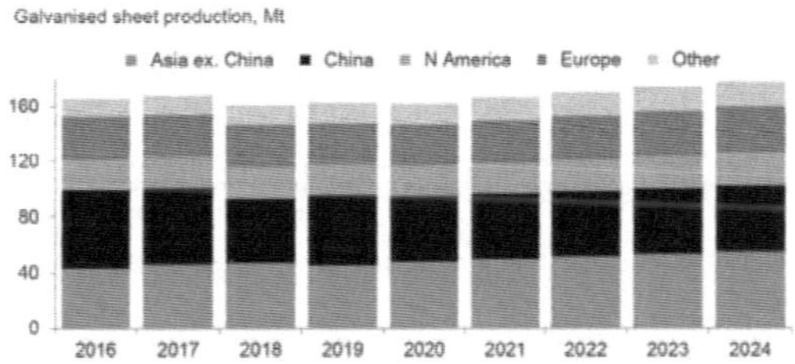

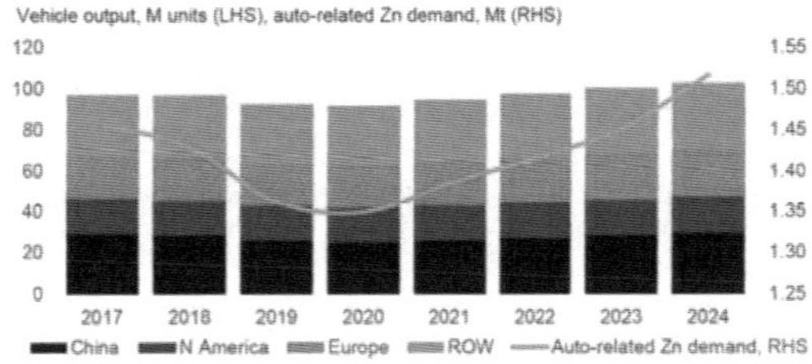

- Galvanised sheet output accounts for around 25% of global demand.
- Output growth over our forecast is expected to come mainly from Asia ex-China.
- Automotive zinc demand accounts for around 10% of the global total but took a hit in 2019, partly due to new emissions regulations in Europe, China and India.
- Global vehicle output is not expected to return to growth until 2021 but should be positive for galvanising and die casting demand going forward.

17

Zinc Insight, February 2020

China's Q1 refined demand has plummeted

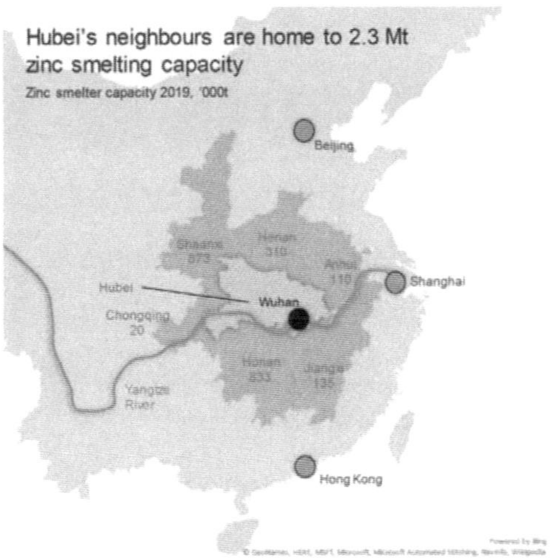

- We estimate that China's Q1 demand will fall by 16% y/y.
- Many manufacturers in provinces where restrictions have been lifted have been slow to restart.
- Most construction sites will not reopen until 21 February at the earliest.
- There is no zinc smelting or mining in Hubei province but neighbouring provinces are home to 2.3 Mt zinc smelting capacity
- Weak demand for zinc and sulphuric acid has already led to output cuts at some smelters.
- More may be forced to reduce output or carry out maintenance.

18

CRU

- Key features of the Zn market in 2019
- Mine supply
- Smelter output
- Demand
- **Market balance and prices**

Surpluses will begin to build from 2020

Zinc market balance, '000 tonnes contained zinc

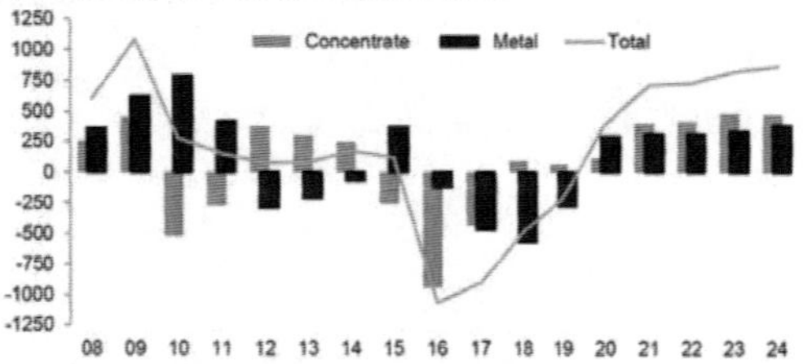

Cumulative zinc market surplus; adjusted; million tonnes

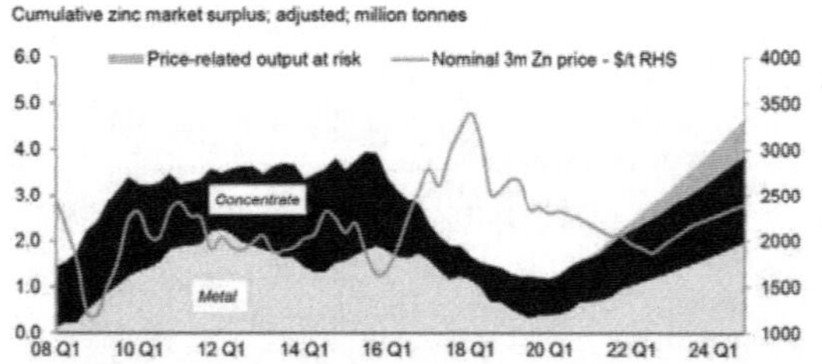

- Concentrate surpluses have been slower to build than expected but TCs remain high, confirming good availability.
- We expect them to remain modest this year but to build strongly from 2020.
- This time last year we expected the refined market to return to balance in 2019. We downgraded demand, but downgraded smelter output by more.
- We now expect a rebound in smelter output growth against still-weak demand to lead to sizeable metal surpluses over our forecast.
- We still believe that prices will need to cut deep in to the cost curve to rebalance the market.

Helen O'Cleary
Senior Analyst
t +44 20 7903 2308
e helen.ocleary@crugroup.com

Alice Fox
Principal Analyst
t +44 20 7903 2261
e alice.fox@crugroup.com

Christine Meilton
Senior Analyst
t +44 20 7903 2098
e christine.meilton@crugroup.com

Dina Yu
Senior Analyst
t +86 10 6510 2206 ext. 352
e dina.yu@crugroup.com

Brian Peng
Analyst
t +86 10 6510 2206 ext. 235
e brian.peng@crugroup.com

Paul Wiggers de Vries
Senior Analyst
t +61 4056 59456
e paul.wiggersdevries@crugroup.com

CRU International Limited
London | Sydney | Tokyo | Beijing | Shanghai | Singapore | Mumbai | New York | Pittsburgh | Sao Paulo | Santiago

Registered in England No.940750. Registered office: Charles House, 108-110 Finchley Road, London NW3 5JJ

About CRU

CRU's reputation with customers across **mining**, **metals** and **fertilizers** is for integrity, reliability, independence and authority.

CRU's insights are built on a twin commitment to quality primary research and robust, transparent methodologies.

CRU invests in a global team of analysts, the key to gaining a real understanding of critical hard-to-reach markets such as China.

We strive to provide customers with the best service and closest contact – flexible, personal, responsive.

CRU – big enough to deliver, a high quality service, small enough to care about all our customers.

Legal notice

This presentation is private and confidential. It must not be disclosed in whole or in part, directly or indirectly or in any other format without the prior written permission of CRU International Limited.

CRU International Limited's responsibility is solely to its clients and its liability is limited to the amount of the fees actually paid for professional services.

Although reasonable care and diligence has been used in the preparation of this presentation, we do not guarantee the accuracy of any data, assumptions, forecasts or other forward-looking statements. We accept no liability to third parties, howsoever arising.

CRU takes information security seriously and currently holds the UK Government approved Cyber Essentials certification. This certifies that we have the appropriate security controls across our organisation and third party suppliers to protect our information assets. CRU also has a privacy policy in place which explains how we handle personal data on our customers.

北矿力澜
BGRIMM LILAN CONSULTING

China's Coronavirus and Zinc Market Updates

Presented by Joe Singer – Penfold Limited

Feb 2020

BGRIMM Lilan Consulting

BGRIMM - Beijing General Research Institute of Mining & Metallurgy

- a state-owned company
- founded in 1956
- a products & technology supplier in the areas of mining, mineral processing, non-ferrous metallurgy, industrial explosives & blasting, processing equipments and reagents, magnet materials.
- a service provider for resource evaluation, technical audits and diagnostics, due diligence, process development, Environment Impact Assessment (EIA), feasibility studies, engineering design and EP/ EPCM /EPC etc.

BLC

- a share-holding company of BGRIMM
- an internationally renowned metals research consultancy to provide marketing intelligences, cost analysis, consulting services on China's base metal industry.
- Depend on a countrywide network of mines, smelters, refineries, fabricators, downstream end users and industry associations
- Gives clients the insight of China's base metal industry.
- Provides independent and professional in-depth advices and consulting services on assets, companies and markets.

The impacts of the n-Cov outbreak

The novel coronavirus pneumonia (NCP) outbreak overview

the 2019-nCov affection areas in China by Feb 17

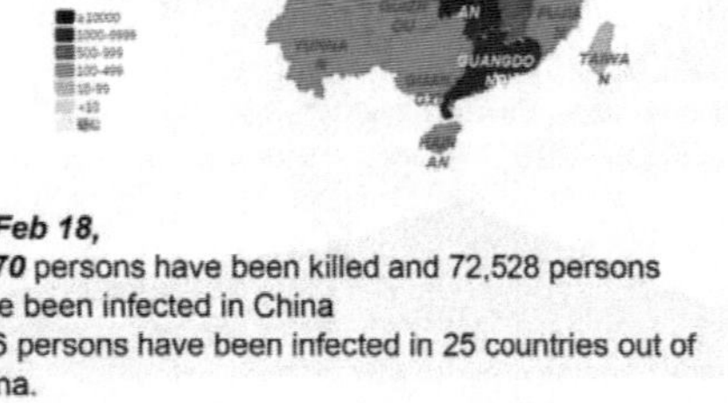

By ***Feb 18,***

- ***1,870*** persons have been killed and 72,528 persons have been infected in China
- 796 persons have been infected in 25 countries out of China.

Dec,2019	The first infection was diagnosed in Wuhan
Jan 23	Chinese government mandated a quarantine of Wuhan
Jan 30	WHO declared the 2019-nCov epidemic as PHEIC
Feb 9	public CNY holiday was extended to Feb 9 at most provinces, 10 days longer than previous years.

new added infected case

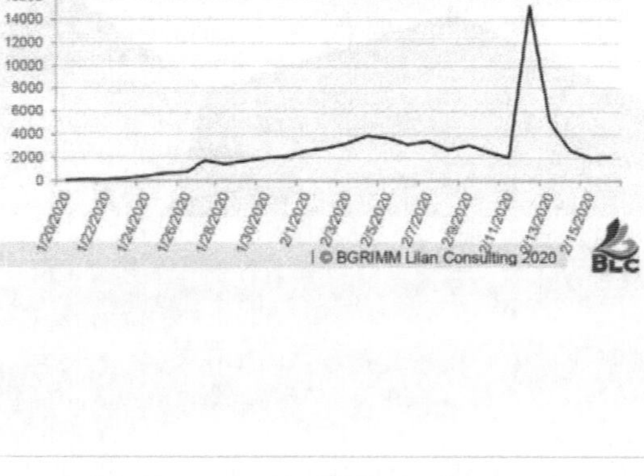

The novel coronavirus pneumonia (NCP) outbreak overview

Statistic of NCP by province/area as of 2020-02-17 13:00

District	Infections	recovery	deaths	District	Infections	recovery	deaths
Hubei	**58182**	**6639**	**1696**	Guangxi	238	51	2
Guangdong	1322	478	4	Yunnan	171	42	0
Henan	1246	455	16	Hainan	163	52	4
Zhejiang	1171	470	0	Guizhou	146	47	1
Hunan	1006	467	3	Shanxi	129	50	0
Anhui	973	266	6	Tianjin	125	45	3
Jiangxi	930	275	1	Liaoning	121	41	1
Jiangsu	626	226	0	Gansu	90	54	2
Chongqing	551	207	5	Jilin	89	30	1
Shandong	541	175	2	Xinjiang	75	7	1
Sichuan	495	136	3	Inner Mongolia	72	8	0
Heilongjaing	457	80	11	Ningxia	70	33	0
Beijing	**381**	**114**	**4**	Hong Kong	57	2	1
Shanghai	332	161	1	Taiwan	20	2	1
Hebei	301	111	3	Qinghai	18	13	0
Fujian	290	84	0	Macao	10	5	0
Shaanxi	240	74	0	Tibet	1	1	0

Source: the National Health Commission

The NCP outbreak impacts on China's economic

Impacts on three sectors

- **the service sector,** as tourism, catering, entertainment and logistics (in particular for small and medium sized enterprises);
- **the manufacturing sector,** typically in mobile phone, automobile and electronic industries, due to the late production resumption, which will to some extent disrupt the global industrial chains;
- **the trade sector,** owing to potential cuts or even shutdown of airlines, sea routes and borders by some countries

Impacts on labor market

- China had around 300 million migrant workers who traditionally return to their hometown for the CNY holiday.
- Hubei sends migrants to Guangdong, Zhejiang, Jiangsu, Beijing, Shanghai etc
- Restriction on travelling and putting travelers in quarantine after the CNY holiday has created labor shortages in labor-intensive industries, including construction and manufactures sectors.

China's GDP might be down by at least 1 to 1.5 percentage points for 2020

- In the short term, such industries as, hospital construction related products, medical apparatus, including thermometer, medical ventilator, robots and related devices for the intelligent medical care solutions, electronic commerce, online game and entertainment are benefitting.
- Central bank is ramping up multiple financial measures, including credit easing and liquidity injection, to stabilize the economy.
- Stimulus efforts forthcoming.

The NCP outbreak impacts on China's zinc production

Not much impact on zinc mine production yet

- Most big mines reported stable production iduring the CNY holiday
- Some small mines in northern China deferred production resumption from Feb 3 to Feb 10.
- The confirmed infected cases in major big Zn mine production provinces of the north and western provinces, such as Inner Mongolia (32%), Yunnan(14%), Shaanxi (7%), Gansu (6.7%), have been low while the infected cases in Hunan province is relatively high.
- Most domestic miners supply concentrate to local smelters has not been impacted by the cross province transportation control.

Impacts on zinc smelter production

- ***Imported concentrate supply*** has been impacted by the inland transportation control, especially for truck transportation. This status are expected to recover by the end of Feb.
- **Acid transportation and consumption**
  - Hubei is the largest P-fertilizer production province, accounting for about 20% of China's sulfur acid demand. Smelter off-gas acid demand has been restricted by production suspension of most P-fertilizer producers in Hubei.
  - Increasing acid stocks, might force some smelters, especially in the provinces, Henan, Hunan, Anhui, to lower production rate if this status could not be reversed within two weeks.

The NCP outbreak impacts on China's zinc production

Zinc first usage

- Most big state-owned galvanized sheet producers remained stable production during the CNY holiday, while smaller ones deferred production resumption for one week.
  - We have seen the increased stocks of galvanized sheet inventories for the weak downstream consumption.
- Some zinc die casters and zinc oxide producers resumed production on 10th of Feb, but production rate is low for weak demand and worker shortage.
- Brass production in Q1 might be down by 25% with low production rate at most brass mills.
- Q1 zinc demand might decrease by 10% YoY while we might see a pick up in O2.

Zinc end-user demand

- Zinc usage in infrastructure and construction segments will be deferred. Some key infrastructure projects have been required to start construction.
- Zinc usage in auto parts in Q1 will be impacted seriously since Hubei is one of major auto parts production bases.
- The epidemic might reduce domestic buying power of the general consumer goods although the subsequent dampening effect on domestic consumption is difficult to be measured at current stage.
- Some downstream producers might lose export orders for the deferred product delivery and travelling restrictions Some countries are looking to diversify supply chains away from China at least in the short term.

The NCP outbreak impacts on China's Zinc Industry

Flowsheet of China's zinc market Unit: kt-Zn

	2019					2020				
	Q1	Q2	Q3	Q4	Yr	Q1	Q2	Q3	Q4	Yr
Zn concentrate production	**901**	**1,130**	**1,145**	**950**	**4,126**	**910**	**1,158**	**1,171**	**971**	**4,210**
Annual growth rate	*-1.5%*	*0.4%*	*1.4%*	*1.4%*	*0.5%*	*1.0%*	*2.5%*	*2.3%*	*2.2%*	*2.0%*
Net import of Zn concentrate	389	310	409	452	1559	402	358	419	458	1637
Other Zn-containing material*	23	36	28	35	122	30	35	36	40	141
Stock changes	*(14)*	*52*	*102*	*(82)*	*57*	*(67)*	*46*	*120*	*(66)*	*33*
Refined Zn production	**1,455**	**1,550**	**1,606**	**1,646**	**6,257**	**1,520**	**1,651**	**1,655**	**1,672**	**6,498**
Annual growth rate	*-2.4%*	*3.8%*	*12.3%*	*12.4%*	*6.4%*	*4.5%*	*6.5%*	*3.1%*	*1.6%*	*3.9%*
Of which Primary zinc	1267	1360	1413	1451	5491	1348	1440	1441	1469	5698
Secondary zinc	188	190	193	195	766	172	211	214	203	800
Net import of zinc ingot and alloy	159	205	136	130	630	114	136	157	153	559
Refined Zn consumption	**1527**	**1816**	**1813**	**1808**	**6964**	**1235**	**1978**	**1918**	**1855**	**6986**
Annual growth rate	-0.4%	-0.1%	0.4%	2.2%	0.6%	-19.1%	9.0%	5.8%	2.6%	0.3%
Stock changes	*87*	*(61)*	*(72)*	*(31)*	*(77)*	*399*	*(192)*	*(106)*	*30*	*71*

*Source: BGRIMM Lilan Consulting * Zinc content in Pb, Cu concentrate*

Key changes:

- 2020 Zn mine production growth was revised down from 2.3%(+93kt-Zn) to 2% (+84kt-Zn).
- 2020 refined Zn production growth was revised down from 4.3% (+270kt) to 3.9% (+241kt)
- Q1's refined Zn usage might fall 10.1% YoY albeit Q2's usage might pick up 3.2% YoY, resulting in a lower annual growth of 0.3% (compared to 0.8% of our previous forecast)

Overview of China's zinc market

China's zinc concentrate supply

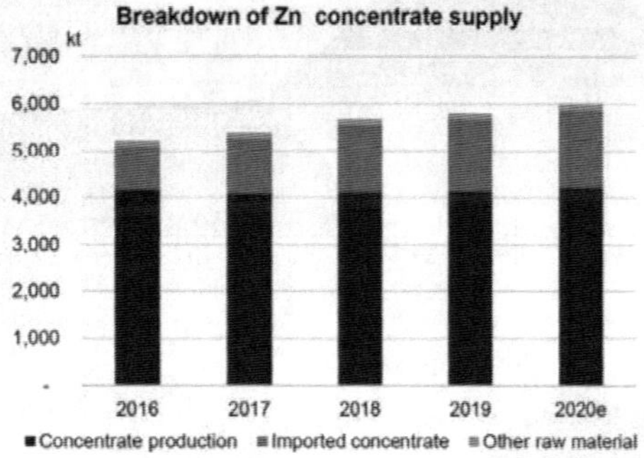

Source: BGRIMM Lilan Consulting

Zn concentrate supply: 2019 + 2.2%(+124kt-Zn) to 5,807kt-Zn 2020 + 3.5%(+202kt-Zn) to 6,009kt-Zn

- Domestic mine production was stable and accounted for 71% of total zinc concentrate demand in 2019.
- Supply increase was mainly from the imported concentrate
- Major zinc mine production provinces: Inner Mongolia(32%), Yunnan(14%), Hunan(7.5%), Shaanxi (7%),Gansu(6.7%)

Zinc mine production

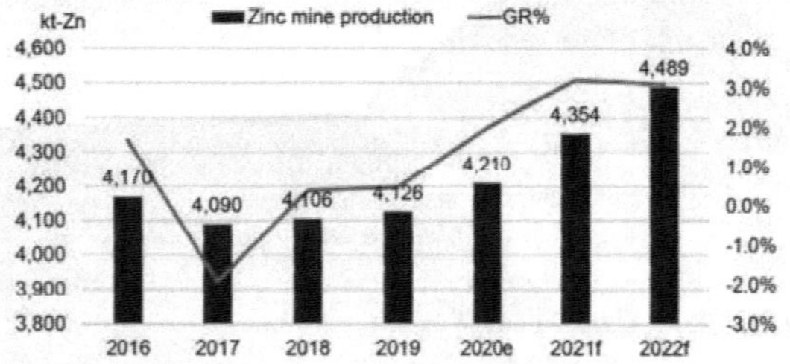

Source: BGRIMM Lilan Consulting

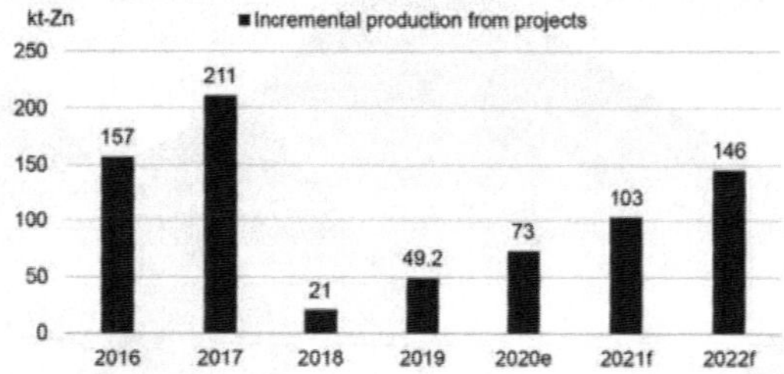

Source: BGRIMM Lilan Consulting

2019 +0.5% YoY (+20kt-Zn) to 4,126kt

Positive factors

- 49kt-Zn increase from the newly commissioned projects
- Production resumptions at small mines in Yunnan and Huayuan(Hunan)

Negative factors

- Safety accidents: Yinman (-30kt), Luoba (-16kt), Nandan(3-4kt)
- Mining right issues : Qinling Mountain
- Later than scheduled production resumptions at some mines

2020 +2.0% YoY (+84kt-Zn) to 4,210kt

Positive factors

- 73kt-Zn increase from new projects and capacity ramp-ups at the mines commissioned in 2019.
- Some mines in Xilin Gol region of Inner Mongolia and Huayuan region of Hunan will increase production in 2020.

Negative factors

- High TCs and falling zinc prices might narrow the profit margin and hurt the operating enthusiasm

Zinc concentrate imports

Zinc concentrate imports 2016-2019

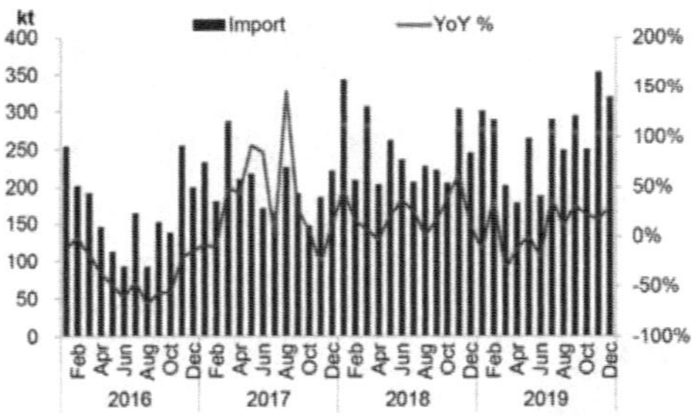

Source: China Customs, BGRIMM Lilan Consulting

Zinc mine production from China invested overseas mines

Source: BGRIMM Lilan Consulting

concentrate imports 2019 +6.9% YoY; 2020 (f) +5.8% YoY

- Chinese invested overseas zinc mines was 768.4kt-Zn in 2019, a 29% YoY up
- Concentrate imports in H2 2019 was up 24% YoY
- High TCs encouraged the smelters to use more imported concentrate (eg. Zhuzhou, Zhongjin Lingnan)

Refined Zn production

Zinc smelter production 2016-2022

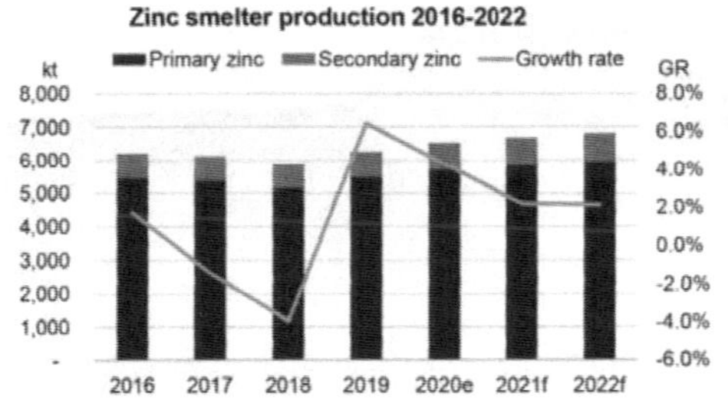

Source: BGRIMM Lilan consulting (BLC)

HW disposal lines invested by Zn smelters

Zinc Smelter	Location	Smelting capacity (kt/a)	HW disposal project		
			Processing capacity (ktpa)	Investment million RMB	Commissioning date
Luoping Zinc&Electricity	Yunnan	100	139.5	191.9	2016
Northwest smelter (Baiyin)	Gansu	315	140	340	June 2019
Western Mining	Qinghai	100	110	52	H1 2019
Hanzhong zinc	Shaanxi	360	250	415	July 2019
Bayannaoer Zijin	Inner Mongolia	220	220	216	July 2019
Tongguan	Anhui	100	82	513	2020
Danxia smelter	Guangdong	115	281	1,584	2021
Total		**1,310**			

Source: BLC collected from public informatior

2019 +6.4% YoY (379kt) to 6,257kt

- Low base of 2018 due to EP issues (Luoping ,Hanzhong, Danxia)
- Refined Zn production has recovered steadily since April
- Zn smelters in Huayuan Hunan which were shutdown for EP issues, resumed production gradually
- Shorter smelter maintenance shutdowns driven by high TCs
- Secondary zinc production should be up 7.5 % YoY in 2019

The Zn smelter leaching residue disposal issues has been basically resolved

- Over 7 HW disposal lines have been constructed or under construction since H2 2018, covering 1.3 million t Zn smelting capacity.
- The average disposal cost is **700~1000 RMB/t-residue** which can be partly offset by the credit from recovered Ag, Pb and Zn.

Refined Zn production

Major zinc smelting projects 2019~2021

	Company	Location	Process	capacity kt/a	Start up	Raw materials
1	Yongxin Nonferrous phase Ⅰ	Guangxi	Extraction	42.5	2019	EAF dust
2	Zijin Zinc	Xinjiang	Conventional	100	2020	Zn concentrate
3	Junlei	Sichuan	Conventional	30	2020	Zn concentrate, etc
4	Fusai Metal New Material	Inner Mongolia	Conventional	32	2020	EAF dust and BF dust
5	Yusheng Nonferrous	Guangxi	Pressure leaching	50	2020	Zn concentrate
6	Yongxin Nonferrous phase Ⅱ	Guangxi	Extraction	57.5	2020	EAF dust
7	Sihuan Zn&Ge (No.3 plant)	Sichuan	Conventional	50	2021	Zn concentrate, etc
8	Yuanfu Zinc	Sichuan	Conventional	50	2021	Zn concentrate, etc
9	Tenglong Recycling	Hebei	Extraction	150	2021	EAF dust and BF dust
	Total			**562**		

Source: BGRIMM Lilan Consulting

Zinc alloy production from zinc smelters

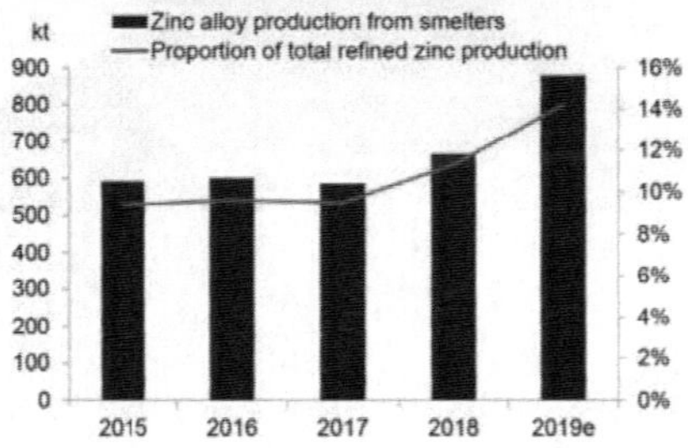

Source: BLC collected from public information

<u>2020 +3.9% YoY (241kt) to 6,498kt</u>

- 40kt production increase from five new projects with 269.5 kt/a capacity.
- No bottleneck of leaching residue issues at the smelter side
- Some smelters (Hanzhong Zinc, Zhuzhou smelter) target higher production for 2020
- Sufficient concentrate supply
- Logistic issues due to the n-ConV-related logistic restriction might result in increasing acid inventories, forcing smelter to cut production

<u>More smelter's zinc alloy production proportion</u>

- Zn alloy production from zinc smelters was up 32% YoY to 879kt, accounting for 14.2% of total refined zinc production in 2019.
- Most Zn producers, such as Huludao Zn, Chihong Qujing, Luoping, Shaanxi Zinc and Yuguang etc, increased Zn alloy proportion in 2019.

Refined Zinc consumption –by first user

Zinc demand break down by first user

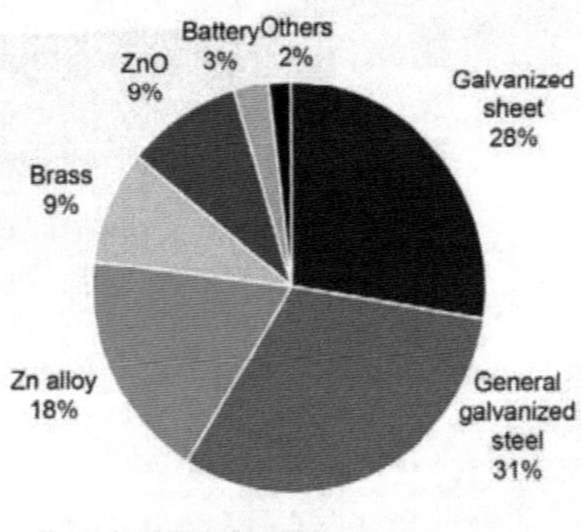

Source: BGRIMM Lilan Consulting

Zinc demand growth rate by first user

	2019	2020(f)
Galvanized sheet	-1.2%	-0.7%
General galvanized steel	2.0%	2.2%
Zn alloy	-0.1%	0.2%
Brass	3.6%	-1.1%
ZnO	-0.3%	-0.6%
Battery	0.2%	-0.2%
Others	-0.9%	-5.9%
Total	**0.6%**	**0.3%**

NCP has impacted Q1's consumption for the late resumption of downstream users

Refined Zinc Consumption –by end user

Zinc demand break down by end user

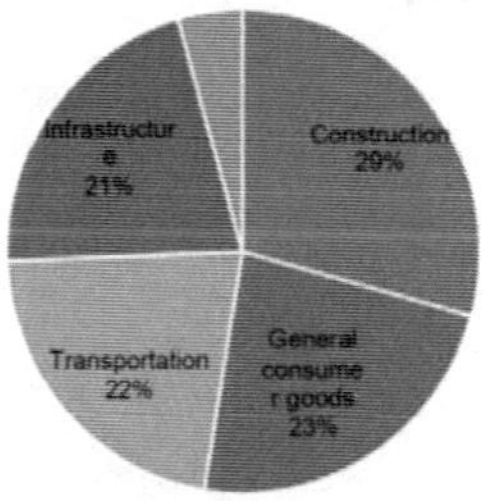

Zinc demand growth rate by end user

	2019	2020(f)
Construction	0.5%	0.5%
General consumer goods	1.5%	-0.5%
Transportation	0.5%	0.5%
Infrastructure	1.5%	2.2%
Industrial equipment and others	1.0%	0.0%
Total	**1.0%**	**0.7%**

Source: BLC

Construction sector flat in 2019 and 2020.
- Real estate investment grew 9.9% YoY in 2019
- New housing starts up 8.5% YoY.
- Housing completion up 2.6% YoY

General consumer goods industry
- Production of aircon, washing machine and refrigerator up 6.5%, 9.8% & 8.1%,
- Better-than-expected production in 2019, was partly driven by strong exports in Jan-October

Infrastructure sector
- Power grid spending in 2019 was down by 9.6% YoY while 5G networks construction has been accelerating
- more infrastructure production in the pipe line, including 5G related projects.

Transportation industry
- Strong road related spending in 2019
- Big drop in auto industry
- Flat railway related spending

Uncertainties: The impacts of the n-Cov

Refined zinc consumption –mixed first user sectors

Galvanized sheet- disappointing

2019:
- Production from major steel producers drop 5.1% in 2019. Negative growth was narrowed in Q4
- Due to 7.5% fall in auto production, although 11% YoY up in export rose due to lifted VAT rebate.

2020:
- Galvanized sheet demand for auto and construction sectors might show a recovery increase in 2020 on the back of low base in 2019 while its demand for home appliance sector will be flat given the high base.

2019:
- **Zn die casting**: weak domestic demand and constrained exports.
- **Brass**: Brass rod demand for aicon valves is strong while brass strip demand is soft.
  - More Zn ingots were used by brass mills with less brass scrap availability in Q4
- **Zinc oxide**: 1.9% YoY growth for tire production in 2019, sluggish renovation market.

General galvanized steel- some highlights
- **Zinc usage in galvanized tube** :moderate growth driven by underground utility tunnels and natural gas tube in 2019 and 2020.
- **Zinc galvanized articles demand for zinc-coated pylon:** low demand for power grid pylon while strong demand for 5G related pylon albeit of increased sharing ratio of pylon in 2019 and 2020.
- **Zinc galvanized articles demand for express ways' related** road signs, isolation fence and guard railing etc reported good performance in 2019, supported by a 9% up in road construction spending. It will remain moderate growth in 2020

2020:
- **Zn die casting**:
  - driven by the moderated growth in housing related market.
  - Increasing market shares of the integrated Zn/Zn alloy producers
- **Brass**: down
- **Zinc Oxide**:
  - tyres -1%
  - ceramics flat (the improved domestic demand offset the drop in ceramic exports).

Conclusion and outlook for 2020

Unit:kt-Zn		2019(e)	2020(f)
Zn mine production		**4,126**	**4,210**
	YoY GR	*0.5%*	*2.0%*
Net import of Zn concentrate		1,559	1,637
Mine stock change		*57*	*33*
Refined Zn production		**6,257**	**6,498**
	YoY GR	*6.4%*	*3.9%*
Net import of zinc ingot and alloy		630	559
Refined zinc stock change		*-77*	*76*
Refined Zn consumption		**6,964**	**6,982**
	YoY GR	*0.6%*	*0.3%*

Source: BGRIMM Lilan Consulting

Strong raw material supply

- More new mining projects
- High overseas mine production
- Increased other raw material

Moderate growth in Zn smelter production

- Less production interruptions at smelters for EP inspections
- Continuous high production under high TCs

More uncertainties in downstream demand

- Strong road related spending
- Improved 5G related zinc tower construction
- auto production? General consumer good production? Exports?

Brief description

- Zinc Mine Production
- The status of Chinese zinc mining projects overview
- Zinc Smelter Production
- Refined Zinc Consumption

Frequency and delivery

- This product includes four issues with annual.
- PDF delivery to subscribers' inbox on 31 Jan. 30 Apr. 31 Oct. and 31 Jan. in electric saved file in English.

THANK YOU!

Zinc Team
Email: kongming@lilanconsulting.com
Tel: +86-10-63299745
Wechat:13810960097

Zinc

IZA February 2020

Jonathan Leng

Wood Mackenzie
A Verisk Business

Trusted Intelligence

woodmac.com

Zinc price lower on soft demand and weak sentiment

Trade wars, trade agreements and the coronavirus

$2400/t in H2 2019 to below $2200/t in February 2020 – DESPITE VERY LOW EXCHANGE STOCKS

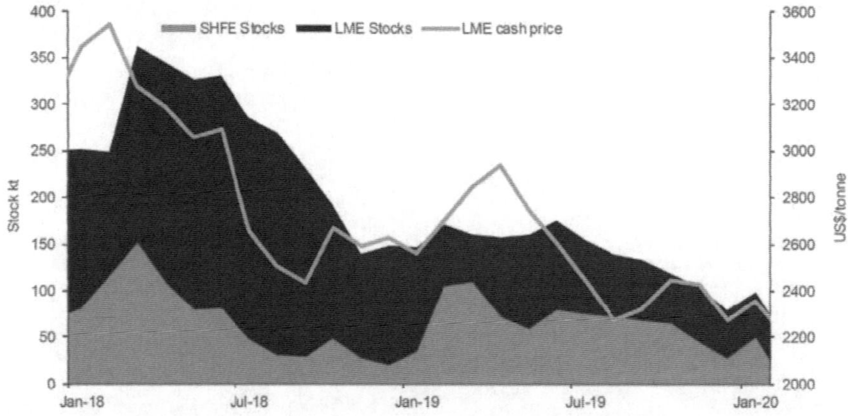

Source: The London Metal Exchange, Shanghai Futures Exchange, Chicago Mercantile Exchange

Modest zinc consumption growth in 2020 follows two years of contraction

Consumption is forecast to grow from 14Mt in 2019 to 15.3Mt in 2025

Changes in zinc consumption by region 2018-2025

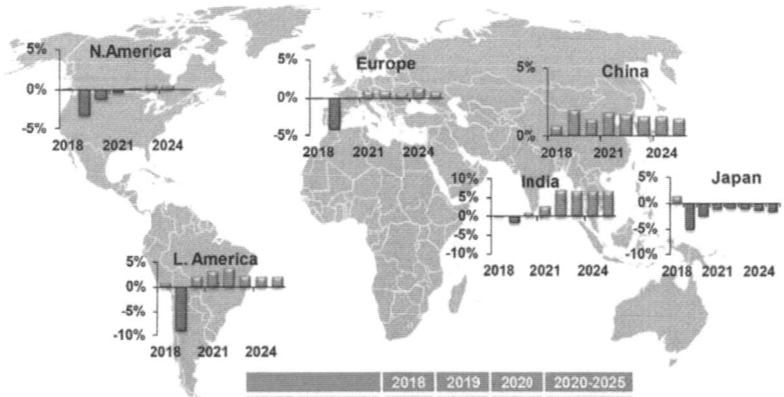

	2018	2019	2020	2020-2025
China	**0.7%**	**1.9%**	**1.3%**	**1.5%**
Rest of World	-1.5%	-3.9%	0.7%	1.6%
Global	**-0.4%**	**-1.1%**	**1.0%**	**1.5%**

Source: Wood Mackenzie

Coronavirus has cut both first- and end-use zinc demand

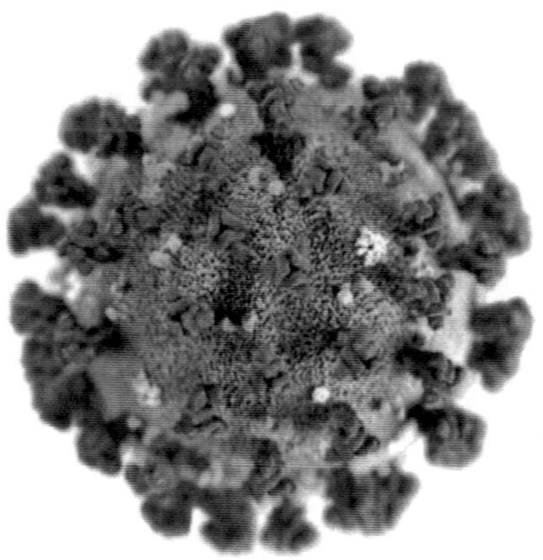

- **Extended shutdowns for galvanizers and manufacturers, especially in southern and coastal provinces**
- **End-use demand depressed as construction work is suspended and consumers unable to purchase white goods or new cars etc**
- **If brought under control in the near future, today's demand weakness will likely be offset by the release of pent-up demand**

Image source: Centers for Disease Control and Prevention

More new mine supply is expected totalling 1.6Mt/a in 2022

New mines and the ramp-up of new projects globally

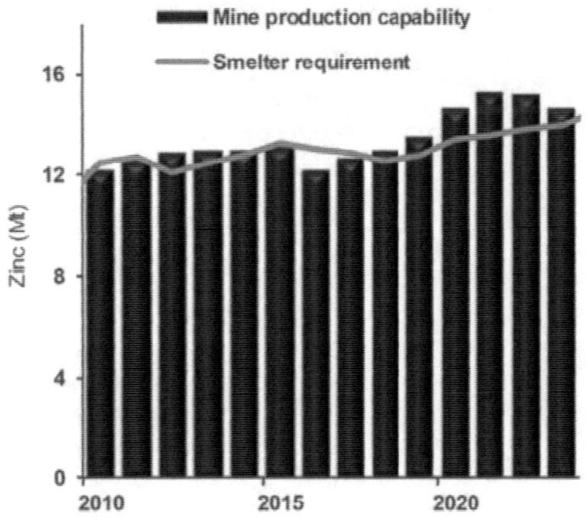

Source: Wood Mackenzie

Ramp-ups		
	Century	+ 100kt Zn
	Gamsberg	+150kt Zn
Expansions		
	Antamina	+100kt Zn
	Changba	+110kt Zn
	Taifeng	+50kt Zn
	Neves Corvo	+85kt Zn
	Penasquito	+110kt Zn
New mines		
	Aripuana	+67kt Zn
	Buenavista	+100kt Zn
	Shalkiya	+125kt Zn
	Zhairem	+160kt Zn
	Others	+450kt Zn

woodmac.com

China cannot fully absorb the rest of world concentrate surplus

Surplus concentrate is in RoW

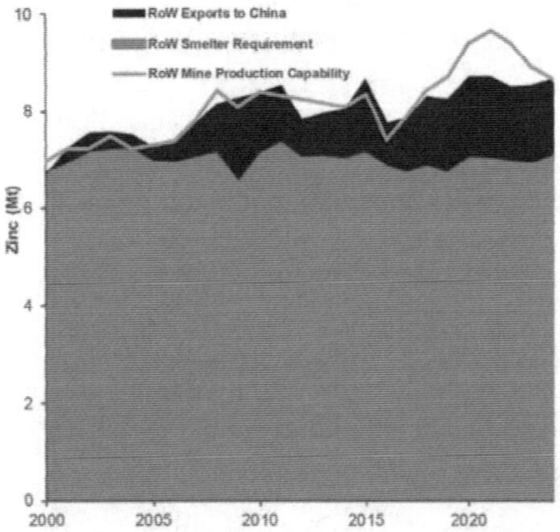

Stricter Chinese import standards – potential risk

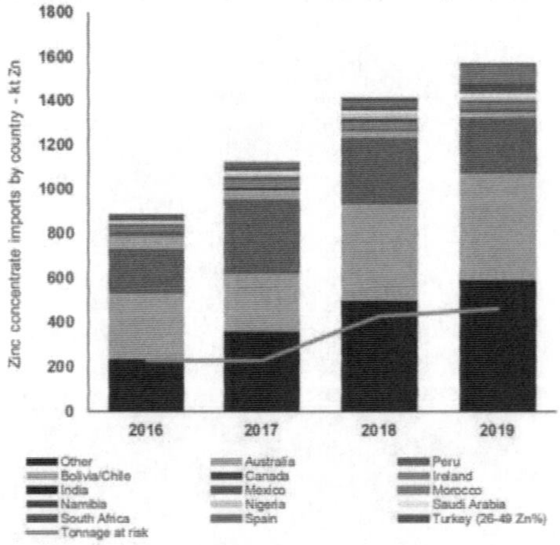

Source Wood Mackenzie, GTT

6

If all planned mine output is produced, the resulting concentrate stocks will be unsustainable

In reality this is highly unlikely to happen due to the effect on concentrate sales and TCs

(UNADJUSTED) the global concentrate balance…

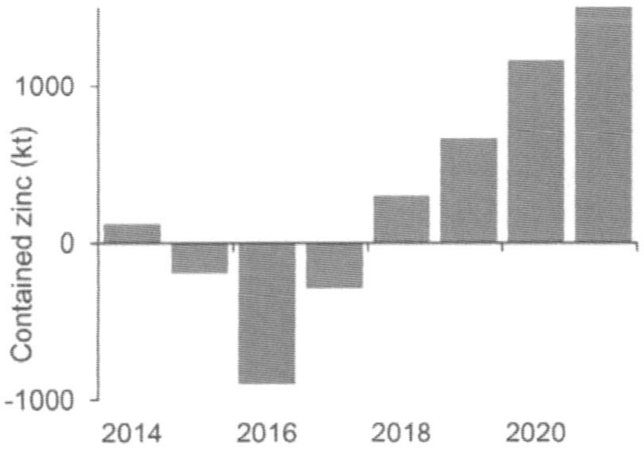

Source: Wood Mackenzie

…would lead to:

Mine cuts and closures

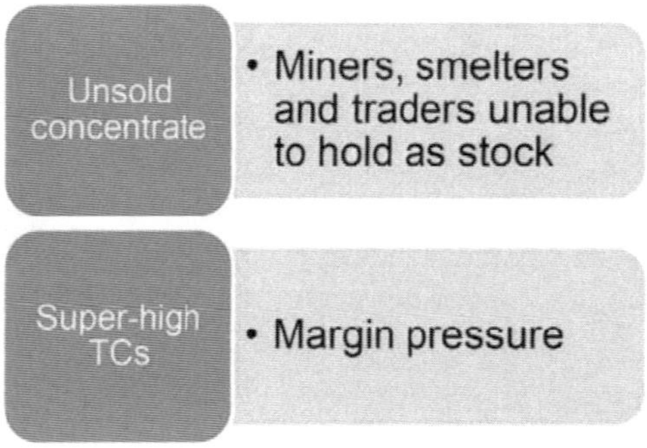

woodmac.com

The zinc concentrate surplus is here…

Even with a large additional disruption allowance to our forecast, we expect concentrate stock to remain elevated in the early-2020's

Large disruption allowance to 2024

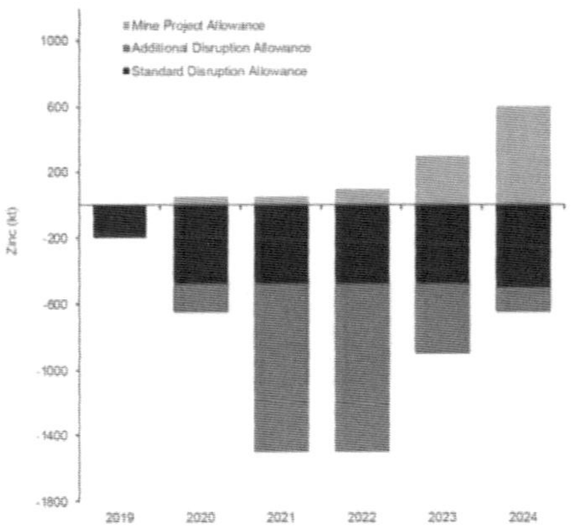

High forecast concentrate stocks and TCs

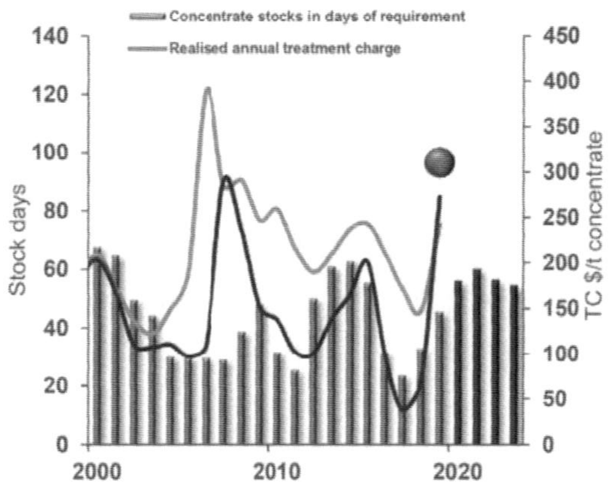

Source: Wood Mackenzie

8

Major risks to the supply forecast

9

Coronavirus is affecting mines and smelters

Novel coronavirus (2019-nCoV) having the greatest impact on miners

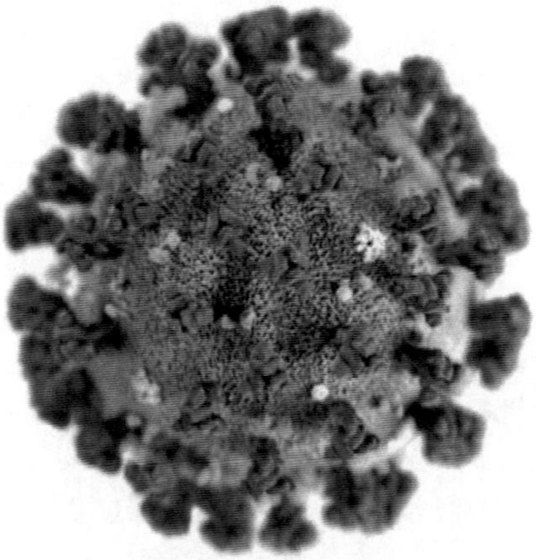

- Mines – Lunar New Year vacation shutdowns extended

- Smelter - acid storage, supply of concentrate and consumables will become an issue

Image source: Centers for Disease Control and Prevention

Significant mine closures typically occur when the price cuts into the 80th centile of the cost curve

M1 margin - cash operating costs

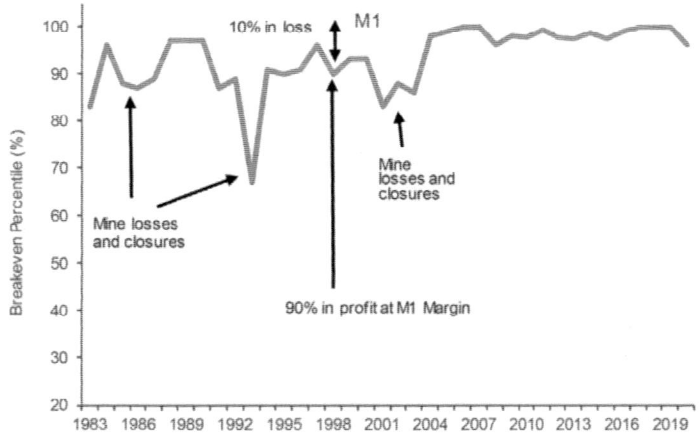

Source Wood Mackenzie

Time to panic as miners' profit margins shrink??

2020 M1 margin curves

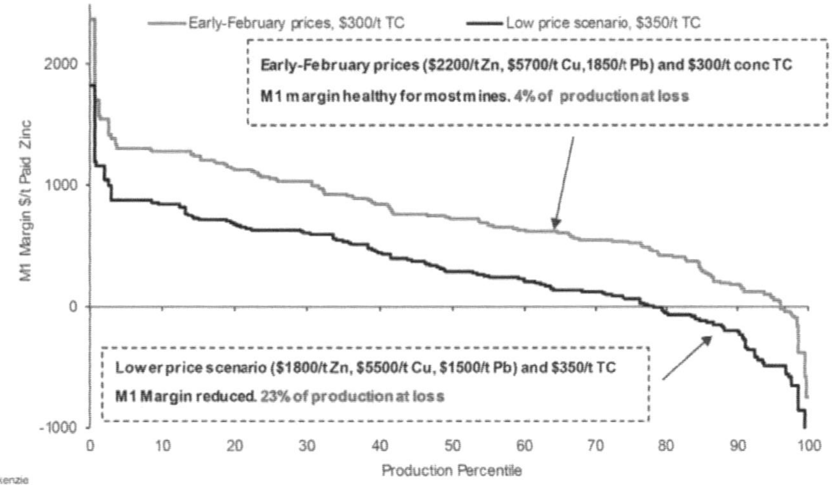

Source Wood Mackenzie

woodmac.com

Zinc mines tend to be polymetallic – diversifying price and TC risk

Sources of revenue zinc-lead mines 2019

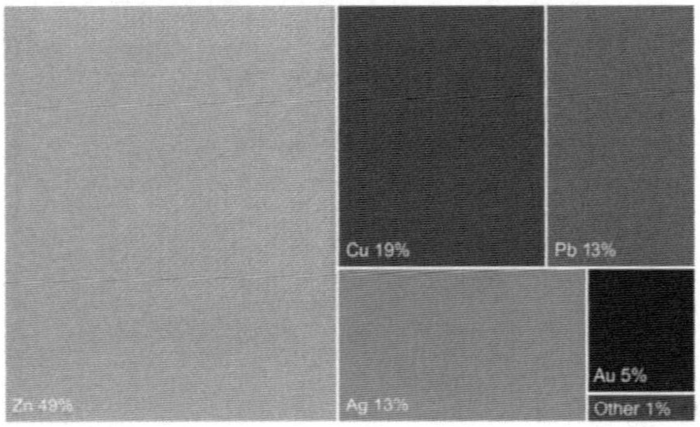

Source Wood Mackenzie

13

woodmac.com

Will plentiful concentrate translate into higher metal production?

Smelter capacity utilisation remains a bottleneck

Utilisation rates are forecast to rise

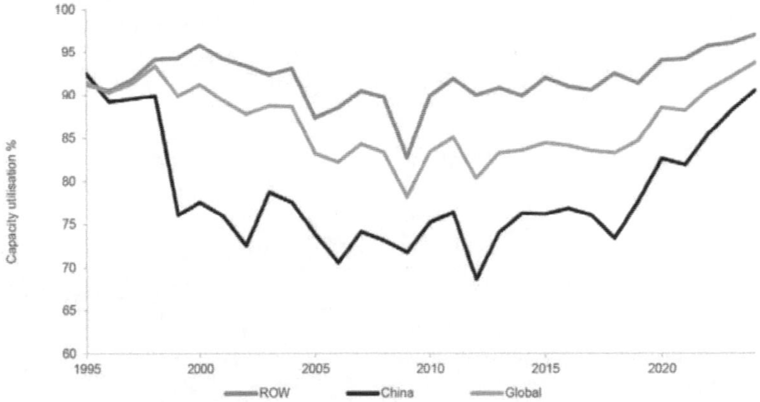

Source Wood Mackenzie

14

Weak demand and market sentiment set to undermine the zinc price in 2020 but low refined stocks to be supportive to 2023

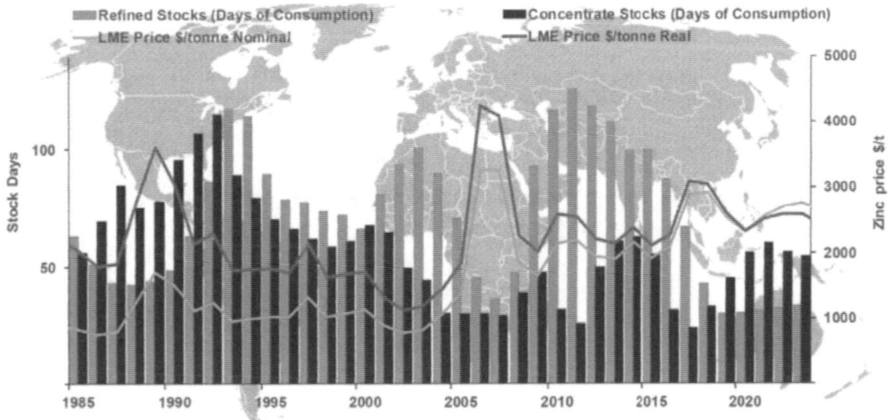

Source Wood Mackenzie

Zinc Key takeaways

- **Sluggish global consumption growth averaging 1.5% p.a. over the next five years**
- **Significant zinc concentrate market surplus in 2020**
  - » Conc stocks to remain at elevated levels for next few years
  - » Elevated TCs
- **Unsold concentrate, high TCs and low price may drive next supply-side correction**
- **Zinc metal stocks to remain low**
  - » Zinc price forecast ranging $2100-2900: based on the recent price - stock relationship
    - » Soft demand and market sentiment could drive the price lower in the range
    - » Positive sentiment from events such as trade agreements could drive the price higher in the range

Disclaimer

Strictly Private & Confidential

- These materials, including any updates to them, are published by and remain subject to the copyright of the Wood Mackenzie group ("Wood Mackenzie"), or its third-party licensors ("Licensors") as relevant, and are made available to clients of Wood Mackenzie under terms agreed between Wood Mackenzie and those clients. The use of these materials is governed by the terms and conditions of the agreement under which they were provided. The content and conclusions contained are confidential and may not be disclosed to any other person without Wood Mackenzie's prior written permission. Wood Mackenzie makes no warranty or representation about the accuracy or completeness of the information and data contained in these materials, which are provided 'as is'. The opinions expressed in these materials are those of Wood Mackenzie, and do not necessarily represent our Licensors' position or views. Nothing contained in them constitutes an offer to buy or to sell securities, or investment advice. Wood Mackenzie's products do not provide a comprehensive analysis of the financial position or prospects of any company or entity and nothing in any such product should be taken as comment regarding the value of the securities of any entity. If, notwithstanding the foregoing, you or any other person relies upon these materials in any way, Wood Mackenzie does not accept, and hereby disclaims to the extent permitted by law, all liability for any loss and damage suffered arising in connection with such reliance.

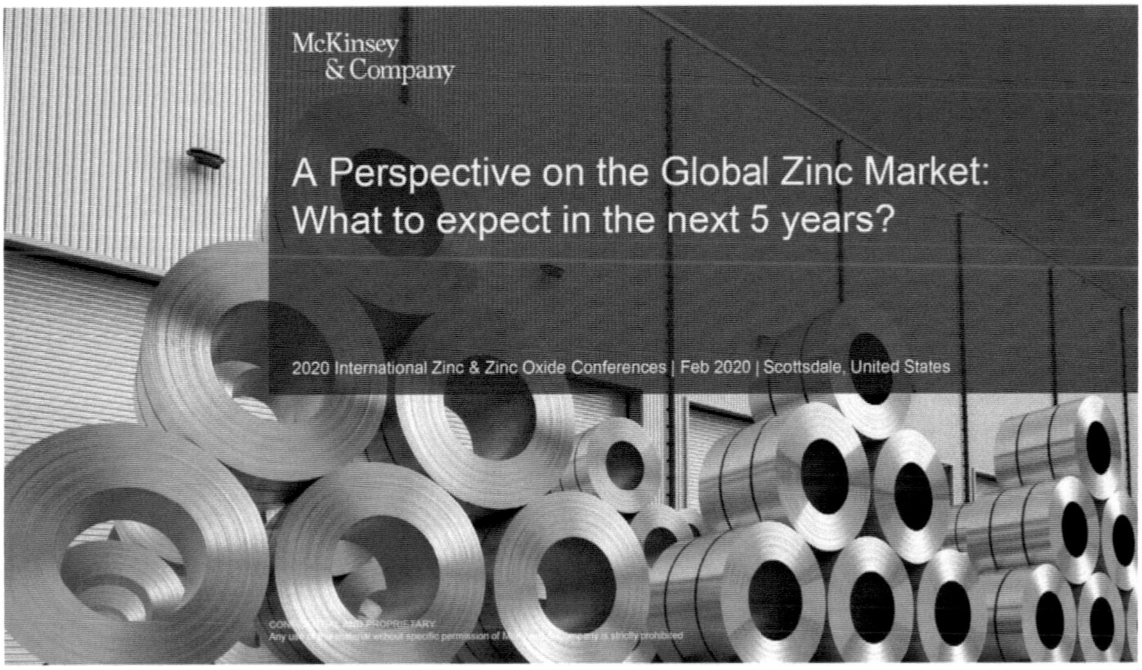

The McKinsey Zinc team present today

Speaker

Ken Hoffman

Senior Client Development Advisor

- Nearly 30 years covering global metals and mining markets
- Leads global assumptions team for all base metals
- Global leader EV battery materials team

Carolina Prates

Senior Knowledge Analyst

- With McKinsey's Basic Materials Institute for over 4 years
- Lead researcher at McKinsey's Zinc Commodity Research Group
- Specializes in Zinc and Steel value chain

McKinsey & Company 2

Agenda

A perspective on the Global Zinc Market

Impacts of Climate Change on Zinc

McKinsey & Company 3

Global zinc market perspective – key messages

Demand

1 Zinc demand is expected to grow at a lower pace than historically at 1.1% p.a. from 2018-23. This is driven by GDP and steel demand growth in China, which remain solid but are decreasing in pace (at 5.7% and 0.8% p.a. respectively)

2 As China's strategy shifts to pursue high-quality developments, rather than only focusing on economic growth rate, we expect a slow down in the infrastructure and construction sectors. Consumer goods, transport, and industrial machinery will drive most of the growth through 2028 in this new, consumer-led, economy

Supply

3 Sustained high zinc prices in 2017-19 have led to a strong mine supply growth outlook of 2.2% p.a. in the next 5 years, which is expected to outpace demand by 0.3 Mt by 2023

4 High treatment charges are being an important driver to incentivize smelters' capacity increases in China, reverting a bottleneck situation that was previously anticipated

5 Share of secondary zinc is expected to increase trough 2023 driven by an increase of availability aligned with new regulations that incentivize recycling

Market balance and pricing

6 Over the last years, the zinc market has been cyclical. After almost four years of structural deficits, the market is reaching equilibrium in 2019-20 and is likely to remain in a fragile oversupply through 2024. It is fragile because the closure of cancellation of a few projects could tighten the market again. By 2025 the lines are expected to cross again, but such expectation should incentivize miners to close the gap by investing in new projects

7 The link to fundamentals like C90 and S-D balance point towards a low greenfield price regime and therefore a decreasing price trend in the next 5 years, ranging between USD 2,000-2,500/t by 2023

McKinsey & Company 4

① In the next 10 years, global zinc demand is expected to deaccelerate as global economies slow down

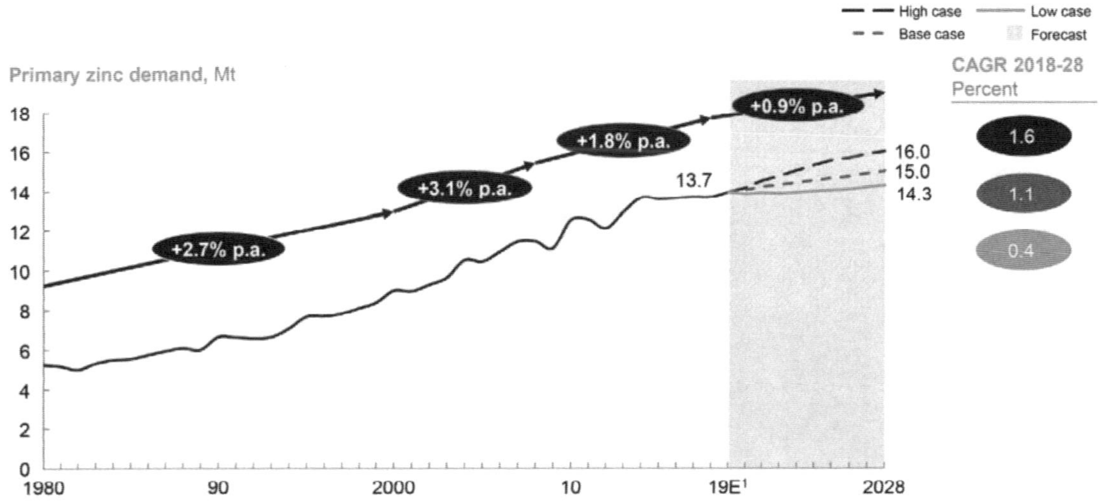

1 Numbers estimated based on actual Jan-Nov numbers

SOURCE: McKinsey Zinc Demand Model; MGI; team analysis

McKinsey & Company 5

① Moderate demand growth should be driven by slowdown in China which still represents ~50% of global demand

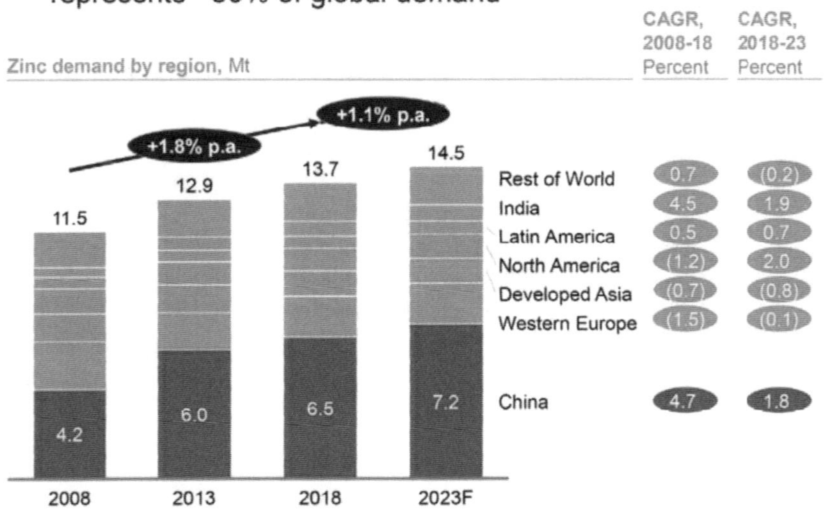

- Chinese steel demand growth is expected to slow down as it has reaches peak steel in 2020 (0.8% p.a. until 2023 vs. 5.5% p.a. 2008-18)
- Chinese GDP growth will remain high, but decrease compared to previous periods (5.7% p.a. until 2023 vs. 7.8% p.a. 2008-18)

SOURCE: McKinsey Zinc Demand Model

McKinsey & Company 6

② In China, zinc consumption for consumer goods, transport, and industrial machinery should keep growing, while construction and infrastructure will decline

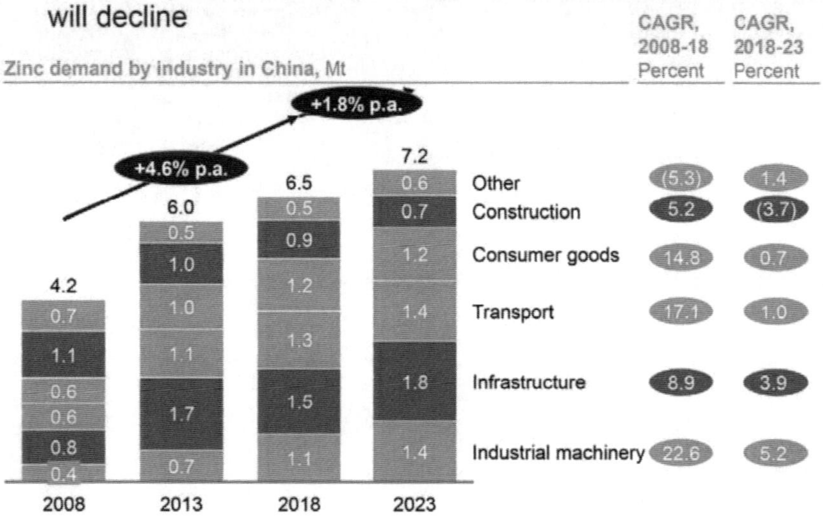

- With the ongoing Chinese economy shift, **we expect to see a decline in the infrastructure and construction** industries
- On the other hand, **industrial machinery, consumer goods and transportation are expected to keep growing** in the future, though at a lower pace than recently observed

SOURCE: McKinsey Zinc Demand Model - China bottom up

McKinsey & Company 7

③ Despite closures and depletion, supply from new projects, restarts and ramp-ups will result in a supply growth of 2% per year by 2023

MineSpans By McKinsey

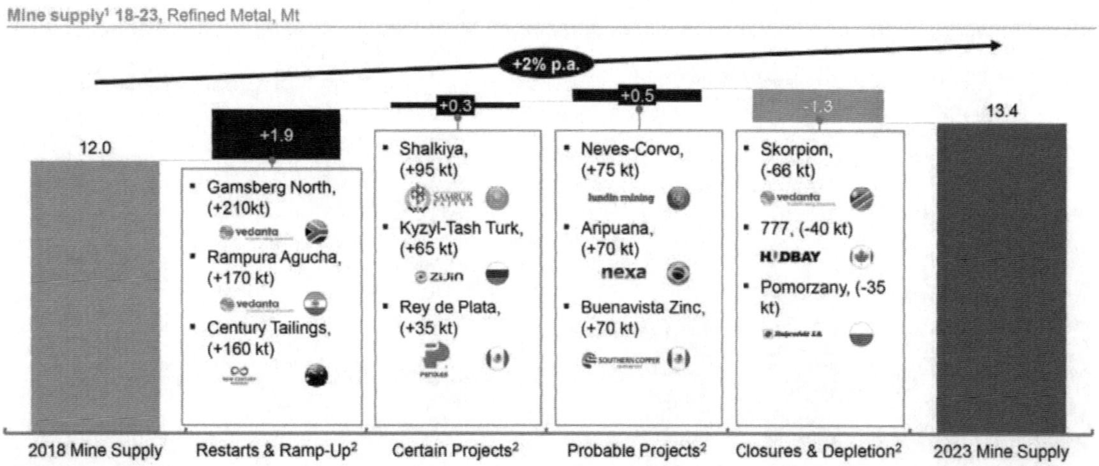

1 Production converted to refined metal basis
2 Categories take into account stage of development, financing stage, location, ownership and economics. Projects which are in construction, in a mining-friendly jurisdiction owned by a major company and do not have excessive capital costs are more likely to be in the base case. Restarts & Ramp-Ups represent the restart of idled operations as well as the ramp up of capacity at recently commissioned projects; Certain projects are typically under constructions with financial guarantees in place; Probable projects are ones with financing complete or announced; Closures & Depletions represent mines that are closing in the future due to depletion as well as other factors.

SOURCE: MineSpans by McKinsey

McKinsey & Company 8

4 On the smelter side, a tight market is expected in 2023, but current pace of ramp ups suggests that production will satisfy demand

MineSpans By McKinsey

Refined zinc demand & supply[1], 18-23, Mt

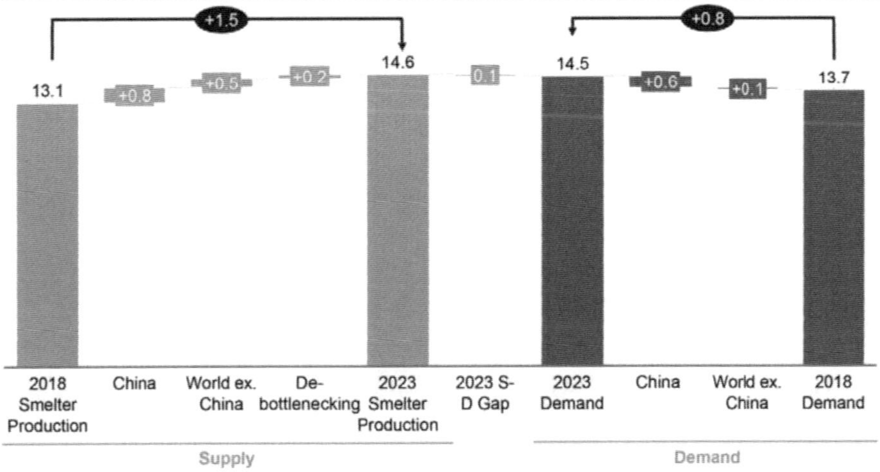

"China's refined zinc output came in at 5.84 million mt for 2019, up 9.64% yoy, as rising treatment charges lifted profits among smelters and encouraged them to operate in high gear."
– *SMM, Jan 2020*

1 Smelting Production
SOURCE: MineSpans by McKinsey

4 Stimulus coming from rising treatment charges encouraged smelters to restart their idled capacities, thus benefiting from increased revenues

MineSpans By McKinsey

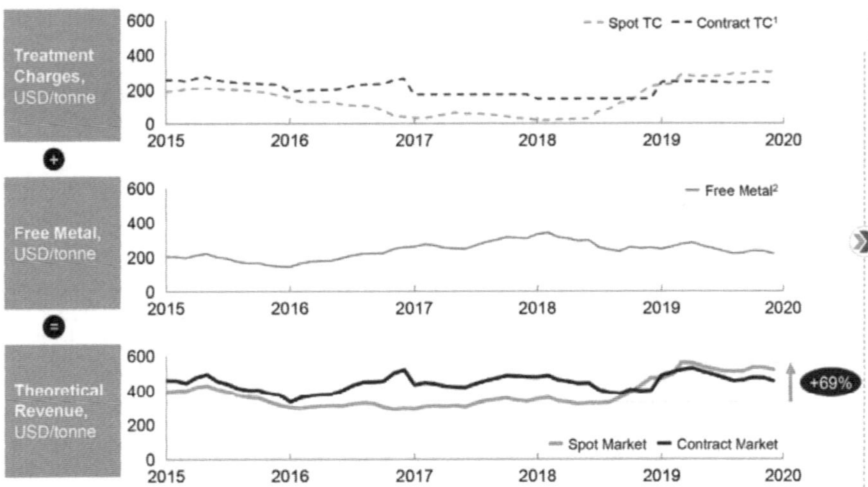

Takeaways

- Companies operating on spot TC could achieve up to 69% higher revenues in 2019 comparing to last year
- Further increase of contract TC is expected in 2020, which should incentivize debottlenecking and expansions

1 Realized TCs after escalation and de-escalation 2 Calculated using LME price, 94.5% smelting recovery rate and 85% payable rate
SOURCE: MineSpans by McKinsey, LME, SMM

5 Share of secondary zinc is expected to grow from 6% in 2013 to 10% by 2023 as more secondary material becomes available

MineSpans By McKinsey

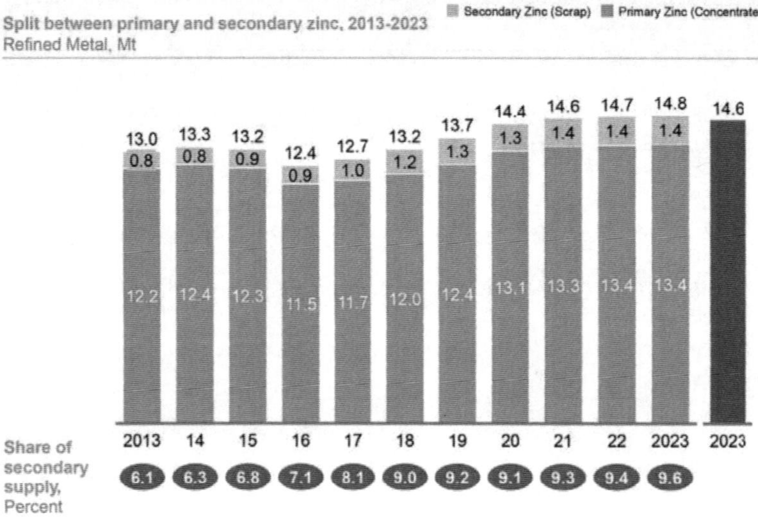

Drivers of secondary supply

- ➕ **Growth of zinc contained in EAF dust in China** as galvanized material consumed in 2000-2005 starts to be scrapped
- ➕ **Governmental incentives support** the recycling industry in China since 2015
- ➕ **Increasing strictness around recycling regulations** and sustainability trends
- ➕ **Increasing demand for the zinc oxide material** due to concerns about concentrate qualities
- ➕ **Steel dust collection fees** are **lower** than **landfill cost**
- ➖ **Upcoming concentrate production growth**

SOURCE: MineSpans by McKinsey, ILZSG, Expert calls

McKinsey & Company 11

6 The mining market is expected to move back to slight surplus in the medium term after years of large structural deficits

MineSpans By McKinsey

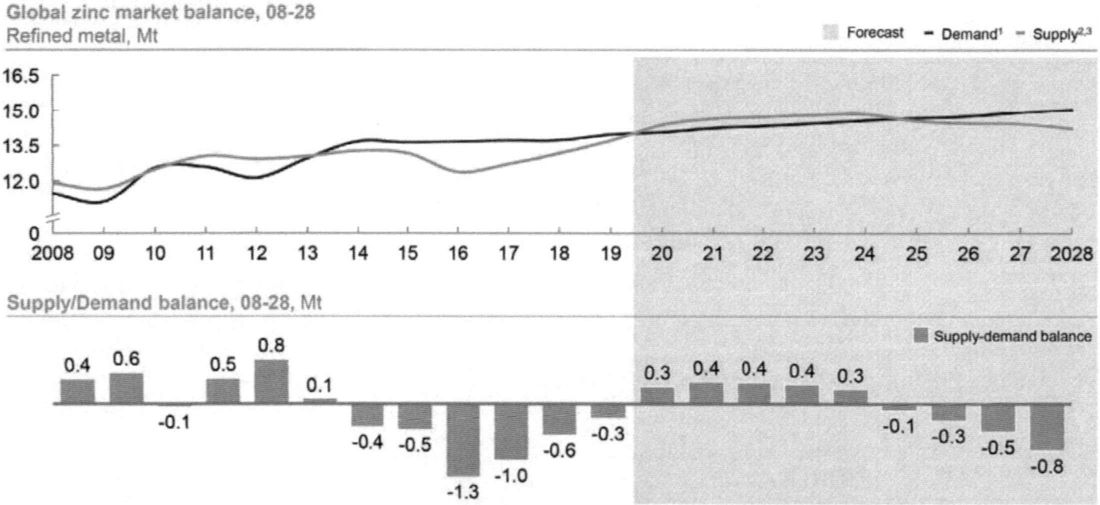

1 Demand scenario: Base case demand 2 Supply scenario: Base case supply including secondary 3 Mining production converted to refined metal basis to make supply and demand numbers comparable

SOURCE: MineSpans by McKinsey, McKinsey Zinc Demand Model, McKinsey Zinc team analysis

McKinsey & Company 12

7 The zinc cost curve is expected to increase between 2018 and 2023 mainly due to increasing TCs and lower lead prices

MineSpans By McKinsey

2018 and 2023 Zinc Mining Cost Curves, total Cash Cost ex. Royalty, USD/t Payable

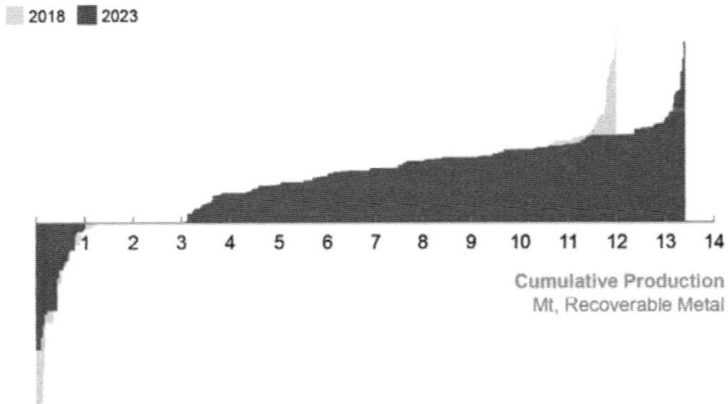

Parameter	Base case	Cost Impact
Brent Crude	↘	▼
Treatment Charge	↗	▲
CNY/USD	↗	▼
INR/USD	↘	▲
PEN/USD	↗	▼
AUD/USD	↗	▼
Copper Price	↘	▲
Silver Price	↗	▼
Lead Price	↘	▲

The fourth quartile mines are operating at a considerably higher cost when compared to 2018 due to higher treatment charges and decreasing lead price. The negative impact of by-product credits is less pronounced in the first quartile of 2023 cost curve.

SOURCE: MineSpans by McKinsey

7 Zinc is expected to move from a 'fly up' to a 'low greenfield' price regime, driven by a fragile oversupply in the mid term

MineSpans By McKinsey

Zinc price and marginal cost to produce[1], 2008-23, US$/t[2,3]

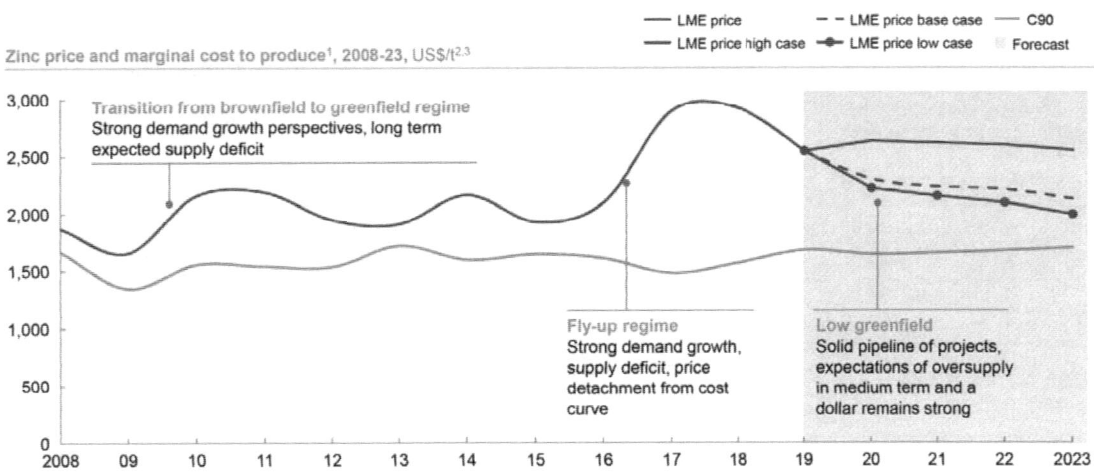

1 90th percentile of cost to produce
2 Historical = nominal, forecast = real
3 Forecasts are based on the assumption that the coronavirus outbreak will be contained soon

SOURCE: McKinsey Zinc price model

Agenda

A perspective on the Global Zinc Market

Impacts of Climate Change on Zinc

McKinsey & Company 15

Major trends related to climate change represent risks and opportunities for the zinc industry

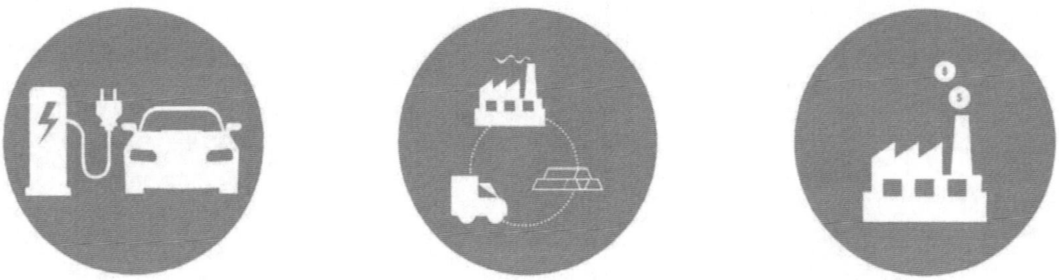

1 Electrification

2 Supply chain disruptions

3 Carbon tax

McKinsey & Company 16

❶ Wind energy is the most relevant for zinc among renewables and likely to impact consumption in the future

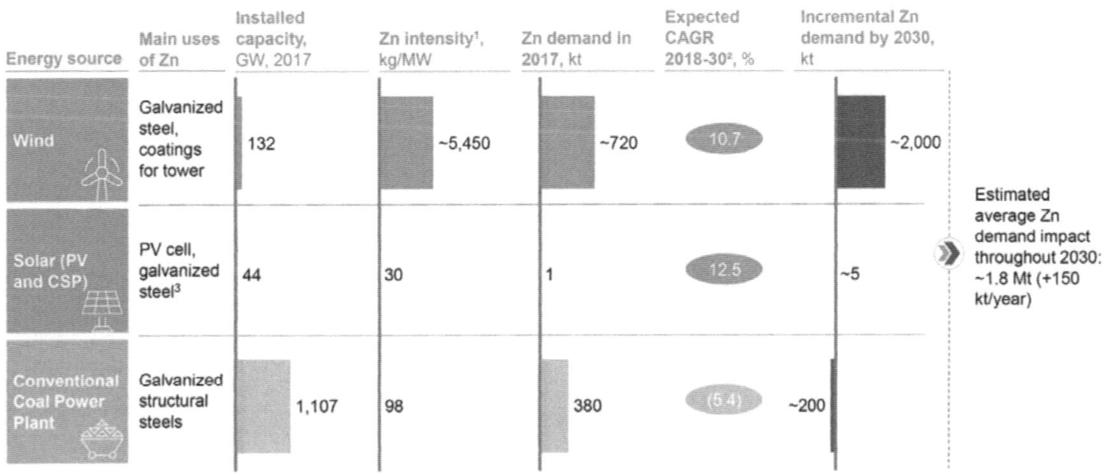

1 Considers only the use of zinc in electricity generation (does not include transmission, distribution, or storage)
2 Assuming IEA 2DS scenario (Sustainable Development Scenario) 2017
3 Zn content in galvanized steel for solar not included in this analysis

SOURCE: IEA Energy Technology Perspectives 2017, World Bank The Growing Role of Minerals and Metals for a Low Carbon Future 2017, press search, team analysis

McKinsey & Company 17

❶ Battery technology is evolving rapidly – Zinc air is not being considered in short to medium term

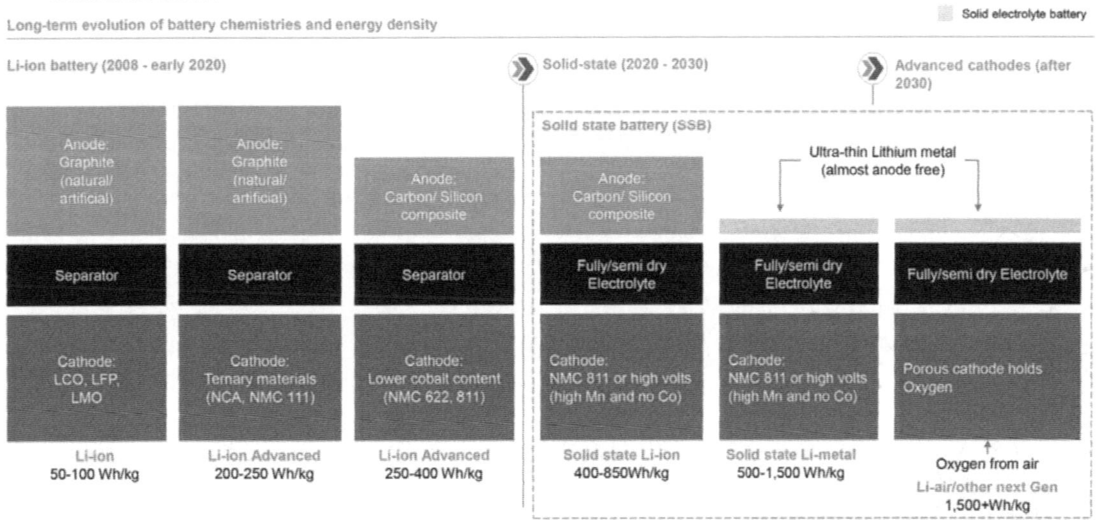

SOURCE: Bernstein, McKinsey

McKinsey & Company 18

2 Nowadays, almost 30% of zinc mining production is located in water stressed areas

MineSpans By McKinsey

2018 map of mining sites against current water stress Water stress level[1], Zinc mines

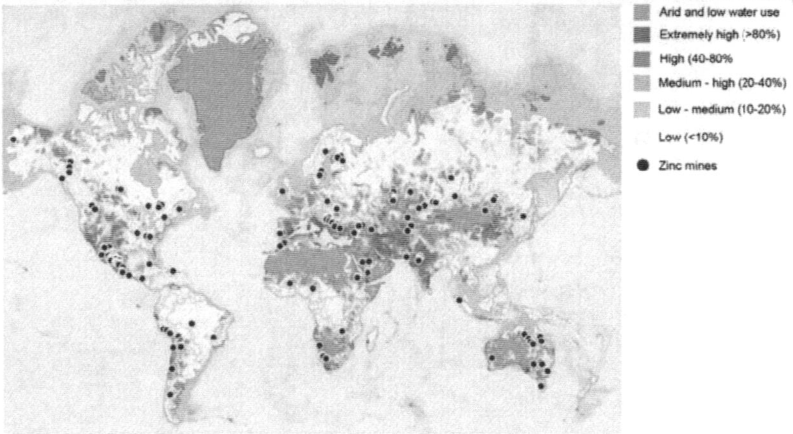

Takeaways

- Total of 3.5 million tonnes (~30% of global production) of zinc is located in areas with high water competition[2]
- Regions impacted are:
  - Western North America
  - Central Asia
  - Western Australia
  - India
  - Southern Africa
- Companies need to reduce water consumption, improve recycling and secure water supply by building new infrastructure

1 Water stress defined as ratio of water use to supply
2 High competition for water is considered to occur in areas where water stress is "High", "Extremely High" and "Arid"

SOURCE: Aqueduct Water Risk Atlas, World Resources Institute (2015), MineSpans

McKinsey & Company 19

2 In 2040, additional 8% of zinc mine supply are expected to switch to high level water stress under current climate change trajectory

MineSpans By McKinsey

Projected change in water stress in 2040 until current climate change trajectory
Water stress level change[1], Zinc mines

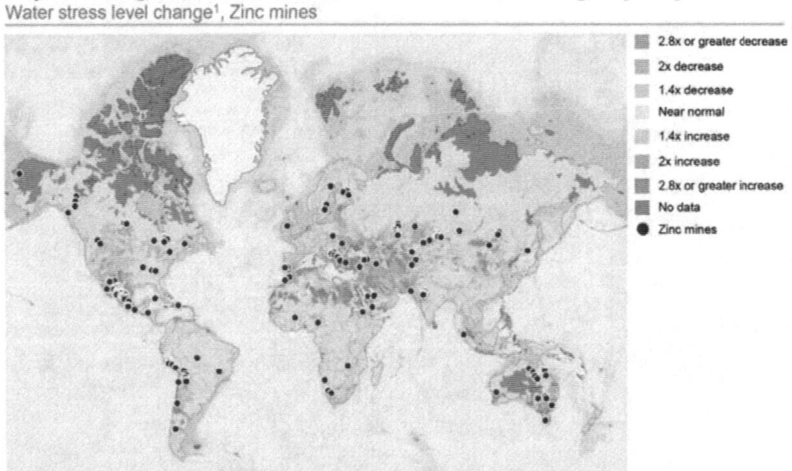

Takeaways

- By 2040, water stress is expected to increase due to:
  - Increased temperature
  - Reduction of rainfall in already water scarce areas
  - Population growth and increased competition for water
- Increased water stress occurs in areas that are already water stressed
- By 2040, regions accounting for ~1.0 million tonnes switches from "medium-high" water stress to "high" water stress
- Main regions impacted will be China and Turkey

1 Water stress defined as ratio of water use to supply. Projected change based on RCP8.5 scenario

SOURCE: Aqueduct Water Risk Atlas, World Resources Institute (2015), MineSpans

McKinsey & Company 20

3 Global CO_2 taxation could lead to higher costs and prices, causing a price increase of ~150 USD/t for every 100 USD/t of CO_2 tax

MineSpans By McKinsey

2018 Zinc Mining Cost Curves assuming potential 100 USD/t CO_2 and price impact, Total Cash Cost ex. Royalty, USD/t Payable

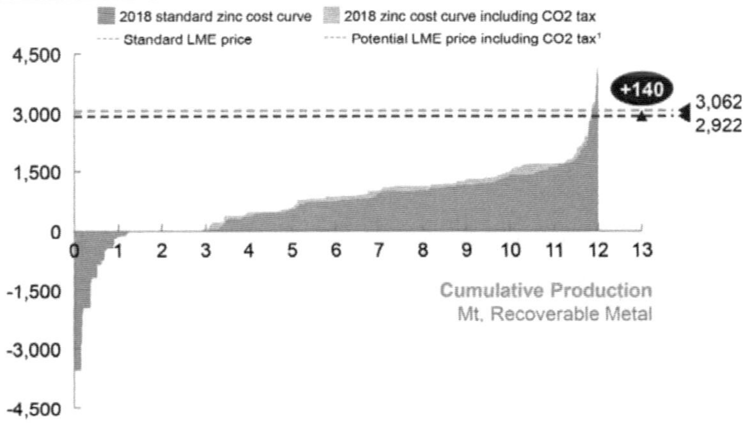

Takeaways

- Countries with high coal depending grids (e.g. China) will be affected the most
- Price increase may incentivize companies to reduce CO_2 emissions through optimization of processes, increased recycling and alternative technologies
- Companies that are successful to reduce CO_2 emission will increase their margin
- There could be a premium on price for consumers who value low carbon products

1 Price with CO_2 taxation defined as 'original price + (CO_2 emission per ton for a marginal player * CO_2 price)'

SOURCE: MineSpans, McKinsey analysis, McKinsey Metals and Mining decarbonanisation model

McKinsey & Company 21

Q&A

Formulating Zinc Oxide in Concrete

Peter Robinson
Integrated Resource Management

IZA Zinc oxide conference, February 19th 2020

Why Concrete?

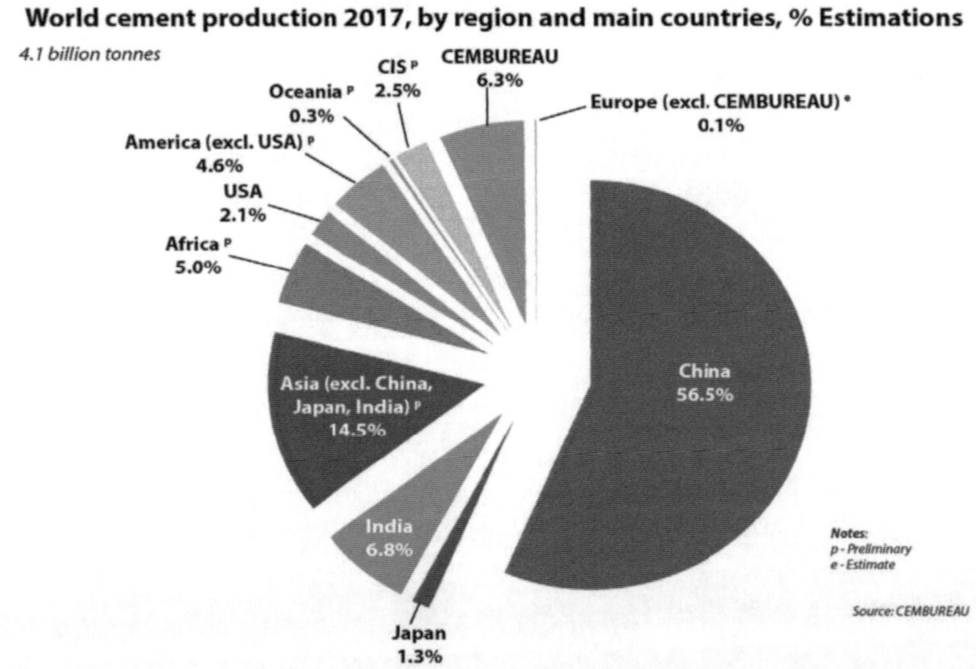

ZnO in Concrete – Initial Work

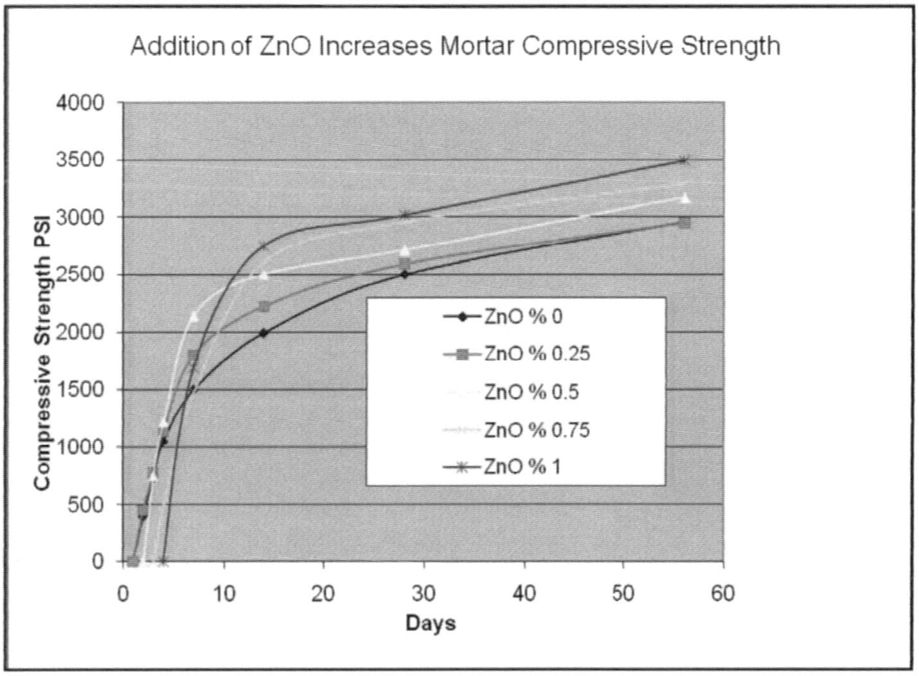

ZnO in Concrete

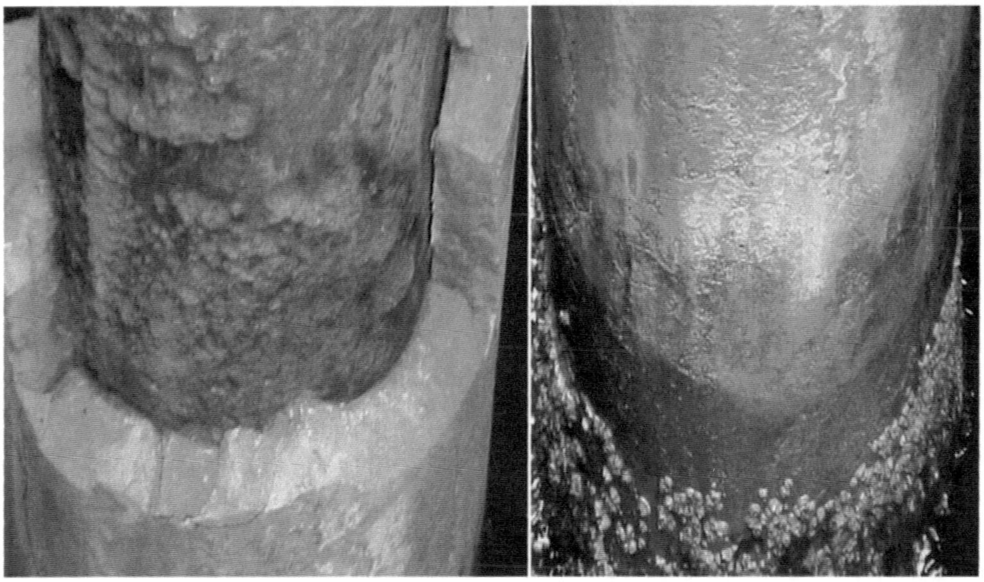

Steel pile protected with concrete, 0% ZnO Steel pile protected with concrete, 1% ZnO

The formulation dilemma

- ZnO is a very powerful retarder, 0.25% relative to cement retards setting for 24+ hours.
- ZnO as an additive to concrete provides the highest level of corrosion protection to embedded steel.
- Corrosion is the ultimate mode of failure of concrete structures – At an enormous cost to infrastructure.
- The ability to formulate ZnO is a key to extending structure lifetime.

Type C Fly Ash

- Fly ash is a (waste) product of coal fired electricity generation.
- Fly ash is major constituent of concrete formulations, due to improved characteristics.
- The calcium content of the ash is dependent on the coal source, high calcium (>15%) is termed Type C and <15% Type F.
- Type F accounts for around 85% of generation.
- India and China account for most fly ash generation. Output in Germany, USA, UK and Canada has fallen in recent years, due to cut backs on coal fired power plants

Montana S.U. work

- A cross section of Type C fly ash generated in the US was evaluated as a sole binder, in place of Portland cement. Approximately 40% appeared to be acceptable.
- Type C fly ash from the Montana Corette facility was used for local projects and discussions led to the incorporation of ZnO in formulations.
- The addition of ZnO to the Corette ash formulations much improved the cementitious properties, in particular the very rapid build in compressive strength.
- The optimum level of ZnO for formulations where the sole binder was fly ash was 8%.

Test data ZnO – Type C Fly Ash

- For example at 2% ZnO, compressive strength was 1800 psi after 12 hours. At 6% ZnO it was 3500 psi after 6 hours.
- At 8% ZnO, compressive strength was 2,500 psi after 7 hours and 6,400 psi after 48 hours.
- At the 8% ZnO addition, the practical advantage was that the rate of build up of compressive strength enabled the installation of interior work surfaces in a single work day, i.e. mix, cast, cure and polish.

Continued work on fly ash…

- Cooperation with Geomatrix ceased.
- Rationale that ZnO would enable the use of lower performing Type C ashes in concrete.
- Corette facility closed and Type C ash was sourced from another facility (showing acceptable results in Montana S. U. work) for IZA tests.
- Test results did not mirror Corette data.
- Fly ash availability became much reduced due to concerns with coal fired energy generation.

Concrete Industry goals

- Reduce dependence on Portland Cement to avoid CO2 generation
- Extend structure lifetime
- Increase use of alternative curing mechanisms, for example alternative concrete technology that utilises CO2
- Increase use of reactive materials that are readily available or derived from waste products, such as ground limestone, glass or steel slag.

Status January 2019

- Test data generated had not confirmed that ZnO could be practically formulated into cement based products.
- We still did not have formal test data that could be used to promote the project to the next stage
- It was understood that ZnO is such a powerful retarder that the cementitious binder has to be more reactive than Portland cement.
- Type C fly ash was no longer considered the preferred material due to availability and uniformity concerns.

2019 Work

- Our previous approach had been to design and evaluate a matrix of laboratory tests at dedicated concrete testing laboratories. This has not been sufficiently flexible to make useful progress.
- Accordingly we have modified our approach to in in-house testing utilising much simplified procedures.
- This has resulted in the ability to direct the research on the fly.

2019 Work objectives

- Previous work has shown that ZnO is the best concrete additive to protect rebar. The major advantage of protecting rebar is that the lifetime of the whole building is extended.
- *According to a recent colloquium at the Getty Center, the average life span of a conventionally built building (masonry and wood) is about 120 years. But for modernist buildings (reinforced concrete and glass curtain wall) it's half that: 60 years.* - **Brandon Graham Donnelly**

- **House bill targets $20B for US bridge repairs**

Strategies to balance ZnO retardation

- The apparent easy fix – concrete accelerators does not work.
- ZnO surface area not expected to be relevant.
- New, more reactive pozzolanic additives may be advantageous. Very fine ground limestone and very fine ground waste glass for example. Such materials have been developed to take the place of fly ash.
- Previous work with calcium aluminate (reactive cements) had not given encouraging results.

ZnO surface area

- If you thought that surface area would not be relevant because concrete is very alkaline with pH typically 12 – 13 and therefore ZnO would all react, you (like me) would be wrong.
- Clearly some ZnO must react for the extreme retardation to occur.

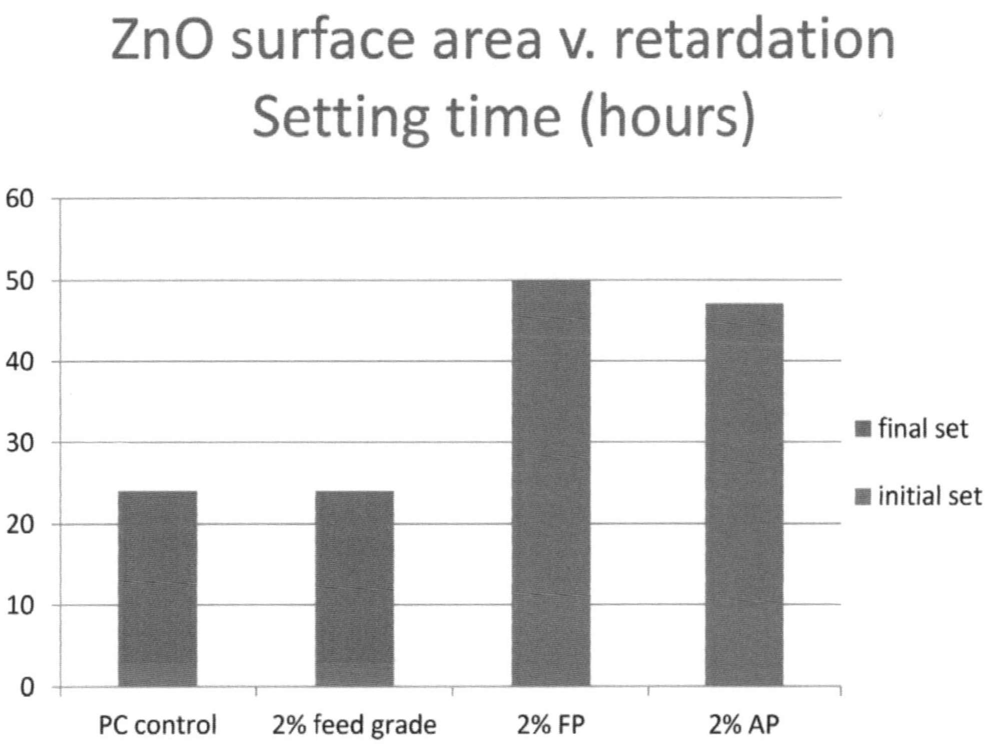

ZnO Surface Area

- Zinc in solution provides highest retardation.
- AP ZnO at 1.5m2g-1 is slightly less retarding than FP ZnO at 5 m2g-1.
- Feed grade ZnO at 0.2 m2g-1 shows zero retardation. - Question – to be confirmed – does it still confer rebar protection?

Reactive Pozzolans

- The reactive pozzolans were evaluated around the 20% Portland cement replacement level.
- Ground limestone - GFP 7
- Ground glass - V 200
- Ground glass - Sioneer
- Both these materials enabled ZnO to be formulated with reasonable initial set times but final curing was still delayed – Workable but not ideal.

Calcium Aluminate Cements

- Calcium aluminate cements have been developed from Ciment Fondu. They are refractory cements used in ZnO furnaces for example.
- These CAC cements react with Portland cement and this route was initially evaluated with Innovative Cements. Results were disappointing. Low levels of ZnO (1%) could be balanced with high levels of CAC (16%).
- This avenue was revisited with collaboration from Kerneos inc., the major N. American manufacturer.
- The reactivity between CAC and Portland cement is illustrated by the Robson Curve.

The Robson Curve

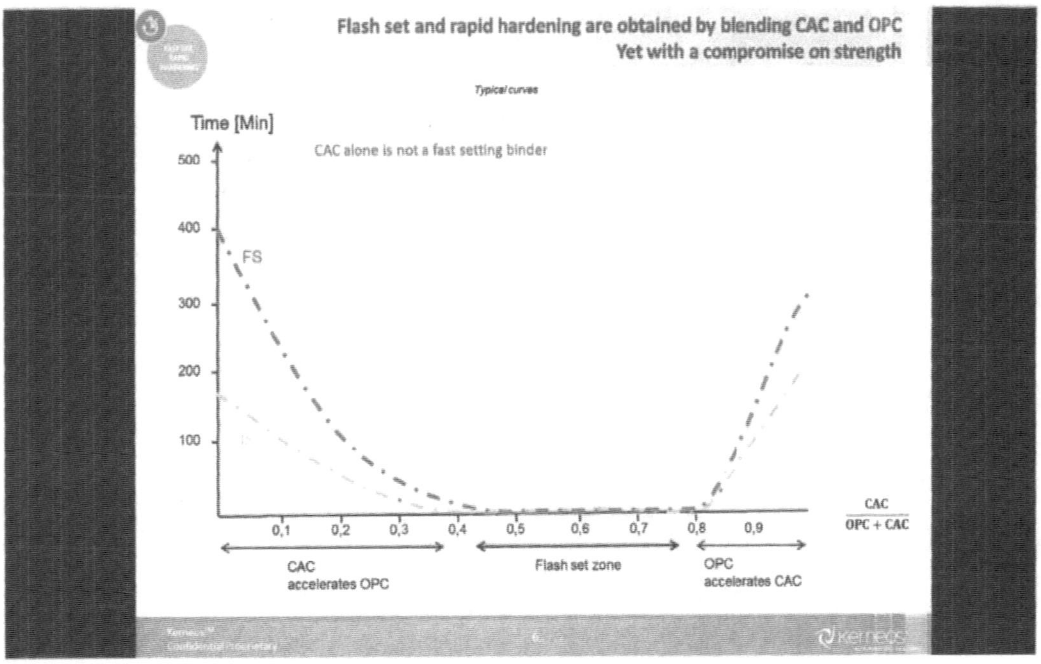

ZnO - Robson Curve

- ZnO does show a strong interaction with the CAC/PC mixture. Contrary to previous work the active point is at a much lower CAC content.
- The ability to counteract ZnO retardation improves as the CAC alumina content increases. CR71>CR51>Ciment Fondu
- The active point is around 4% CAC/96% PC, below this CAC level ZnO is a retarder, above this level it is an accelerator.
- Higher ZnO surface area increases retardation below 4% CAC and increases acceleration above it.
- Higher CAC and ZnO levels appear to allow formulation of accelerated setting and increased early compressive strength (comparable to the Corette fly ash results with 8% ZnO).

Confirmation

- The next stage is to confirm the test results at an recognised testing laboratory, GHD, Waterloo.
- Tests will include-

A. Initial & Final set, Compressive strength build up to 28 days.

B. Accelerated corrosion testing (rapid chloride permeability)

- Mortar formulations to be tested for evaluation-

1. PC
2. Feed grade ZnO/PC
3. CAC/PC/ZnO
4. CAC/PC/ZnO/GF7 (ground limestone)
5. CAC/PC/ZnO/V200 (ground glass)

Products

- Direct addition of ZnO requires a high standard of mixing to ensure uniformity.
- The preferred mode of addition is as a premix with CAC. A 2:1 CAC:ZnO would require an addition rate of 5 - 6% relative to Portland cement content.
- The advantage of this modest ratio is that it is neutral to the characteristics of the cementitious mix.

Costs

- Cement has a modest cost compared to ZnO, around $150 per tonne.
- Performance additives for various uses can be about the same or more than ZnO.
- Ground limestone has lowest cost, c. $50/tonne.
- Ground glass similar to cement.
- CAC ranges from$600 to $1500/tonne.
- Each 1% of ZnO adds around 10% - 15% to the cement cost, depending also on the CAC used.
- Added material cost for 2% ZnO at 25% is modest compared to longer term benefits.

Further Development – for discussion

- Data generation will provide more formal information for publication. Test planned at GHD, Waterloo.
- Publication of findings – ACI
- ACI Foundation (Concrete Research Council)
- Joint work with other suppliers.
- Coordinate work with IZA galvanised rebar project.
- Target projects where rebar corrosion is a significant threat to building/structure lifetime.
- RILEM

Alternative Pozzolan?

'Changes in the pore solution composition and reaction kinetics were measured for cementitious systems with ZnO or sucrose that contained rice straw ash (RSA), wheat straw ash, silica fume, metakaolin, and fly ash. Among the SCMs used, RSA dramatically suppressed ZnO and sucrose retardation'

Authors - Feraidon F.Ataie et al
Concrete Industry Management Program, OCNL Room 229, Campus ZIP 0930, California State University, Chico, CA 95926-0930, United States

Roman Concrete

Setting times (hours)2% FP

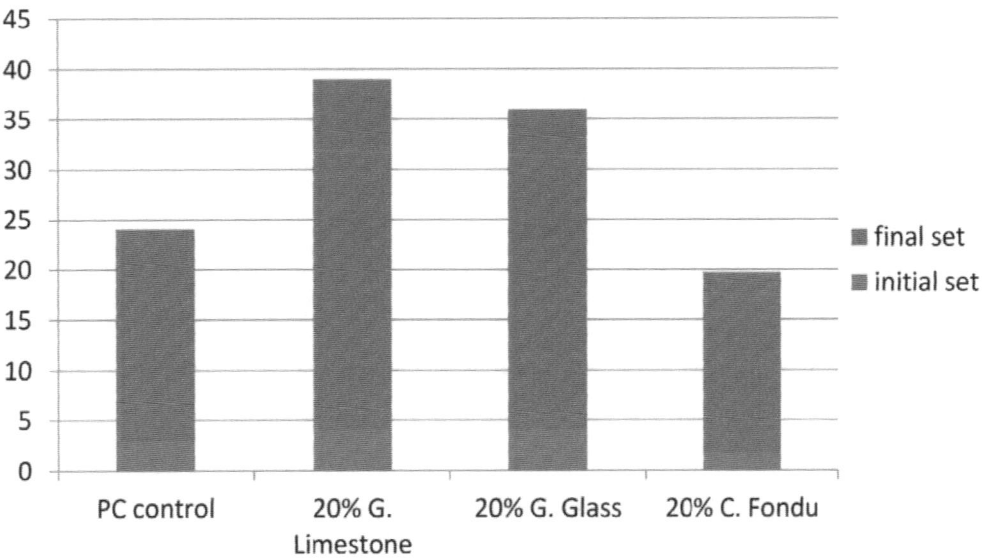

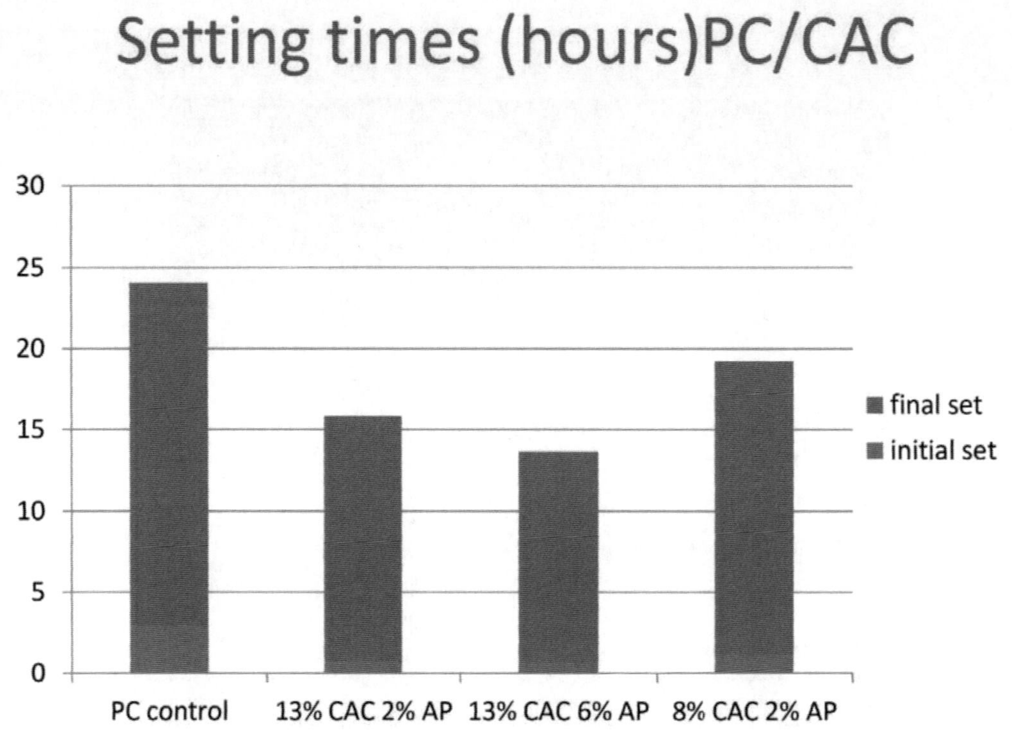
Setting times (hours)PC/CAC
30
25
20
15
10
5
0
PC control
13% CAC 2% AP
13% CAC 6% AP
8% CAC 2% AP
final set
initial set

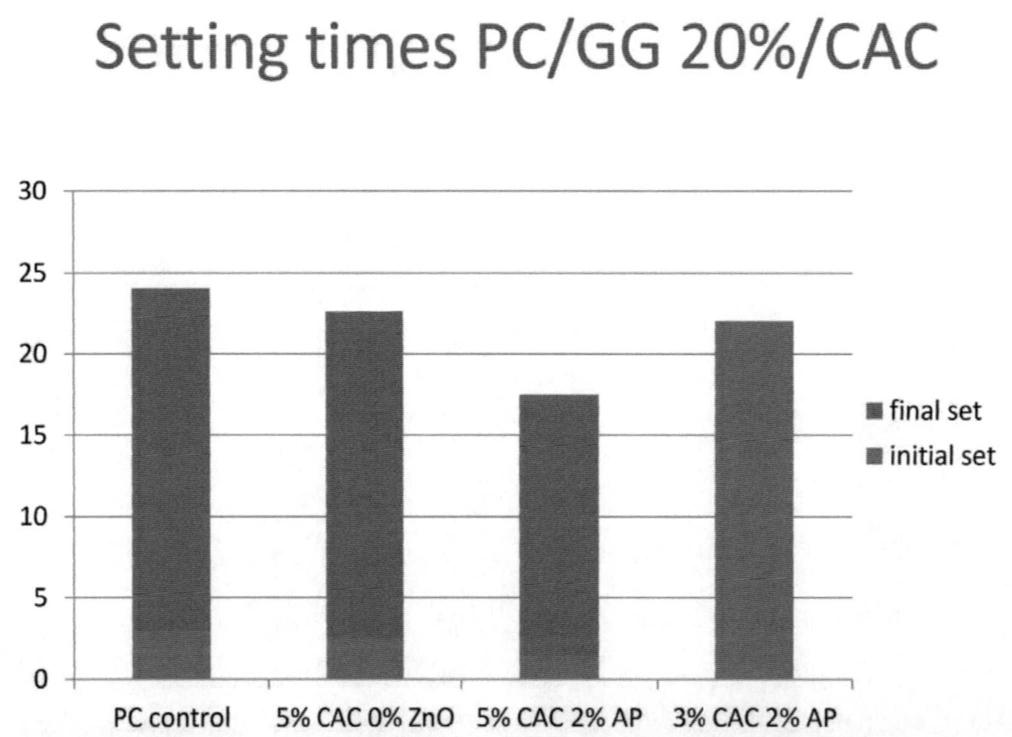
Setting times PC/GG 20%/CAC
30
25
20
15
10
5
0
PC control
5% CAC 0% ZnO
5% CAC 2% AP
3% CAC 2% AP
final set
initial set

Concrete form house - Scottsdale

Thank you

Questions & Answers?

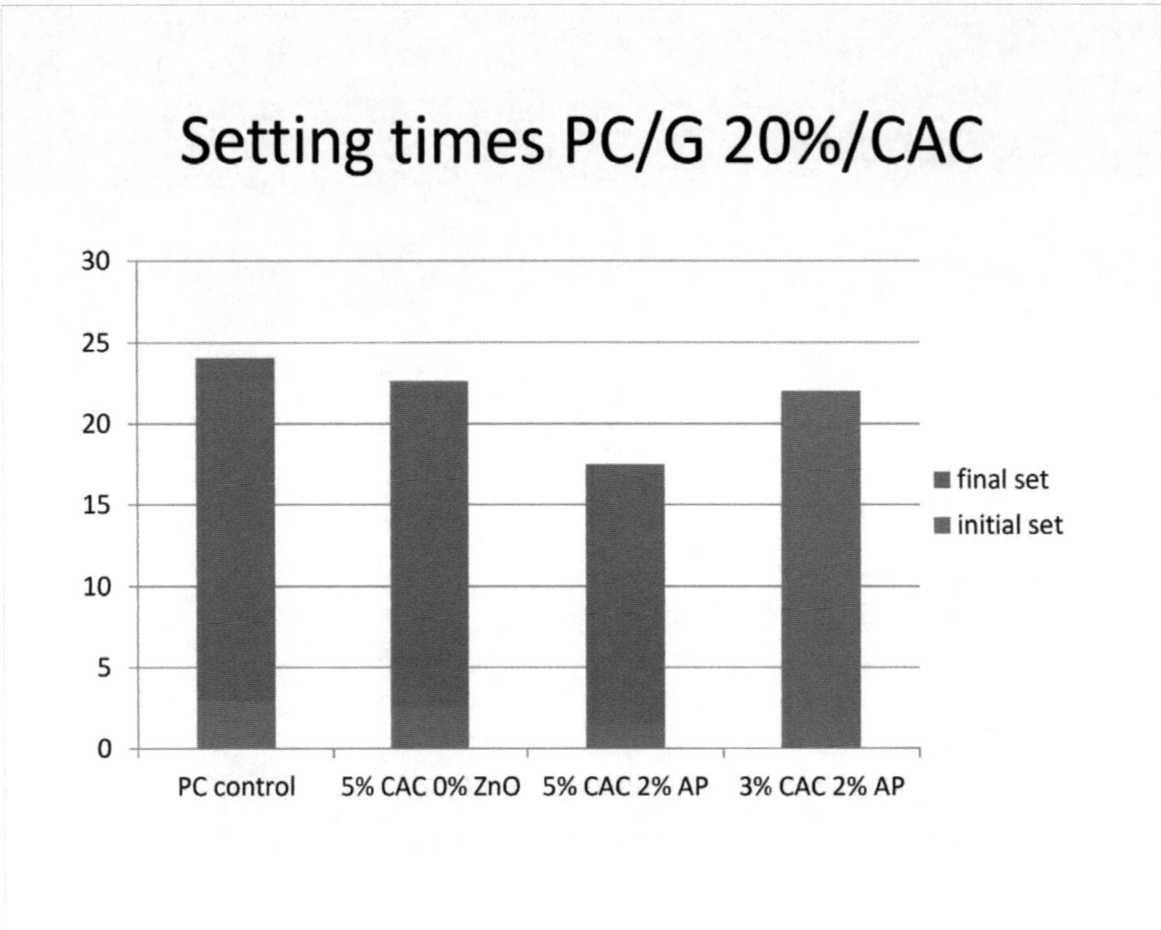
Setting times PC/G 20%/CAC
30
25
20
15
10
5
0
PC control
5% CAC 0% ZnO
5% CAC 2% AP
3% CAC 2% AP
final set
initial set

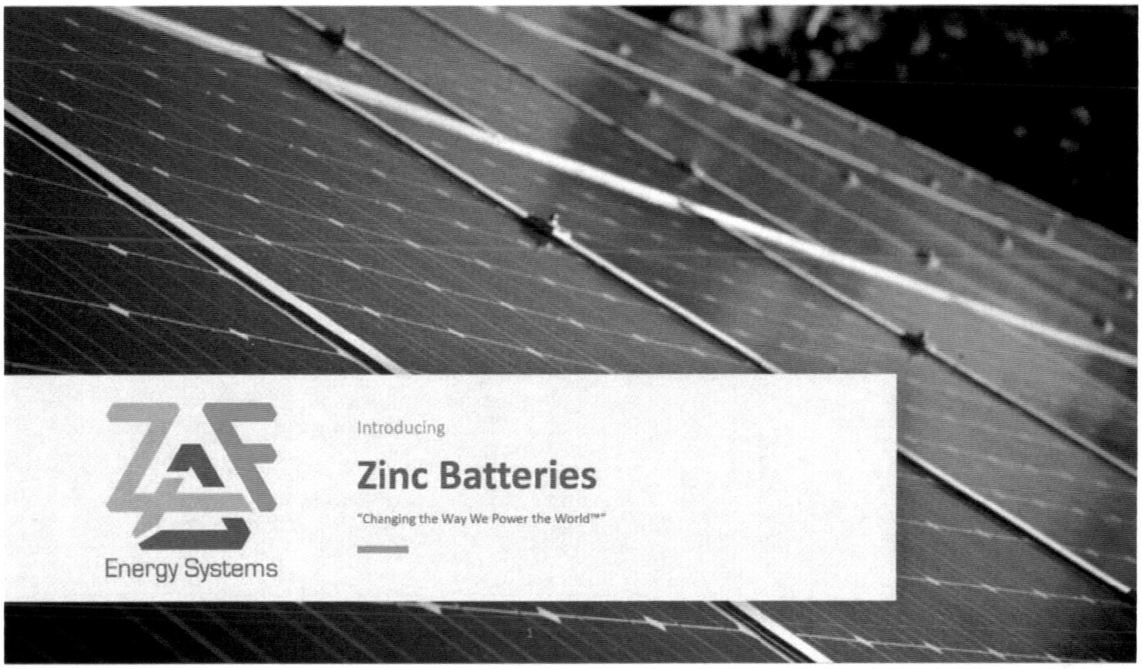

Randy A. Moore

President/CEO, ZAF Energy Systems

Randy has led ZAF for 3½ years on it's journey to commercialize zinc-based batteries.

- Past president of EaglePicher Technologies, LLC, a leading producer of batteries for specialty and government markets. EPT is a $200M company with 1,000 employees and ten locations.
- More than 35 years of senior operational, defense and international experience. Served a business leader at Kollsman, Inc., and Kaman Aerospace, and held management and operations positions at Westinghouse, E-Systems/ Raytheon and Lucent overseeing numerous DoD programs in avionics, aerostructures, electromagnetics, energy storage, radio frequency and acoustic sensors, and communications.
- Retired Lieutenant Colonel in the U.S. Air Force—combined 27 years in the active and reserve Air Force where he served in various capacities including special agent in counterintelligence/ counterespionage for the AF Office of Special Investigations.
- Undergraduate degrees in Marketing and Management at Texas Lutheran University and an MBA from Southern Illinois University.

2

Energy Storage

Unsurpassed Growth

Historic growth rates: 11.66% CAGR Projected

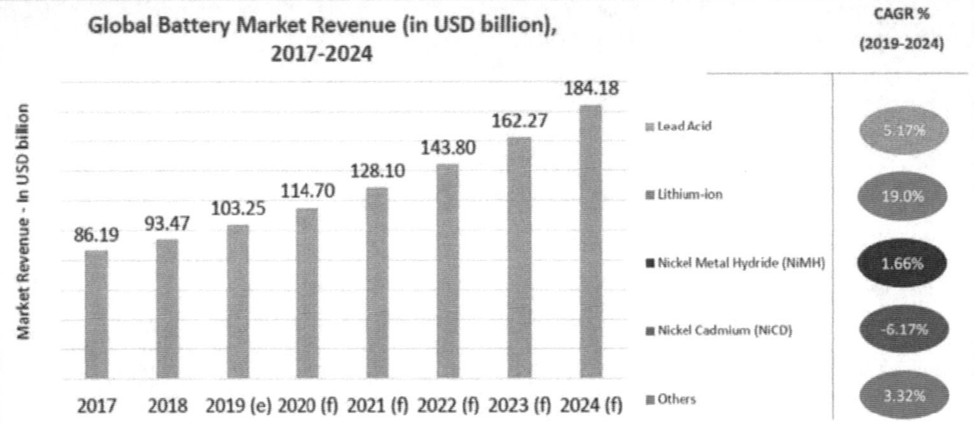

Source: Blackridge Research, Global Battery Market Report, 2019

3

Energy Storage

Lithium Jumps Ahead of Lead Acid for First Time

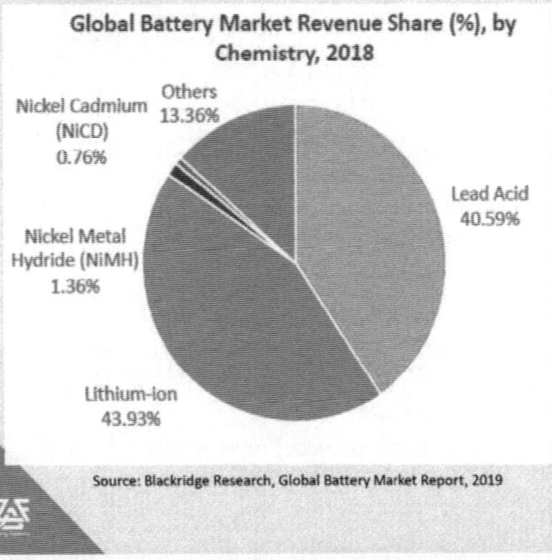

Lithium Jumps Ahead of Lead Acid for First Time

in Billions		2018 Market Opportunity
Chemistry	**Percentage**	
Lithium	43.93%	$ 41.06
Lead Acid	40.59%	$ 37.94
NiMH	1.36%	$ 1.27
NiCD	0.76%	$ 0.71
Others	13.36%	$ 12.49
Total	**100.00%**	**$ 93.47**

Source: Blackridge Research, Global Battery Market Report, 2019

4

Energy Storage

Primary Applications

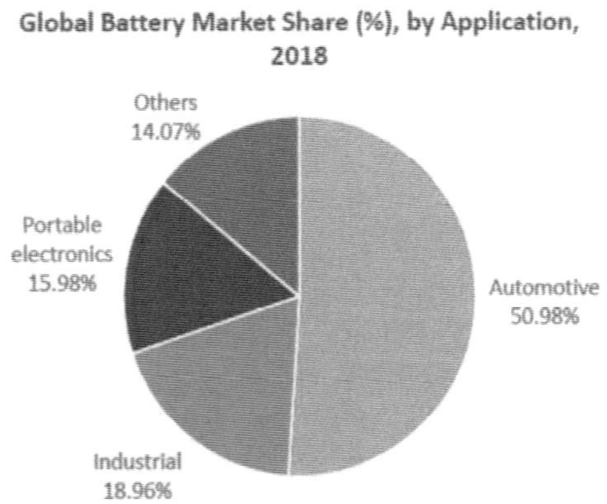

Source: Blackridge Research, Global Battery Market Report, 2019

Automotive Leads the Way

in Billions		**2018 Market**
Application	**Percentage**	**Opportunity**
Automotive	50.98%	$ 47.65
Industrial	18.96%	$ 17.72
Electronics	15.98%	$ 14.94
Others	14.08%	$ 13.16
Total	**100.00%**	**$ 93.47**

Energy Storage

The Dilemma

"Performance Gaps" exposed as technology evolves

Lead-acid Shortcomings

Capacity
Seeking longer run times in smaller packages

Power
More power required for commercial and industrial users

Life
Demand for longer cycle life and shelf life

Lithium-ion Problems

Safety
Customers in key markets place a premium on safety

Complexity
Operations require advanced electronics (BMS)

Cost
Widespread deployment of storage requires < $200/kWh

Market expected to exceed

1 TWh by 2025

Why Nickel Zinc Batteries?

Why ZAF?

ZAF's nickel-zinc battery design solves historic problems.

- ✓ **Electrolyte dry out**
- ✓ **Zinc migration**
- ✓ **Dendrite growth**

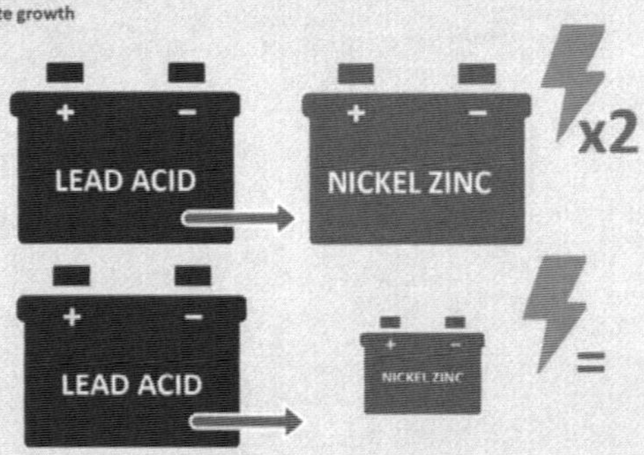

A Nickel Zinc (NiZn) battery can give 2x performance in the same size or the same performance as a Lead Acid battery in half the footprint.

7

Nickel Zinc Group 31

Capacity and Power Comparison versus Equivalent Lead Acid

Battery Data	ZAF G31 NiZn Capacity	Odyssey G31 AGM – Lead Acid	ZAF G31 NiZn Power	Unit
Capacity (C/20)	172	100	152	Ah
Capacity (C/3)	**165**	**84**	**147**	**Ah**
Mass	27	35.3	27	kg
Cycle Life (80% DOD)	700	400	700	Cycles
CCA Current (30s)	650	1150	1000	A
Specific Energy (C/3)	71	27.4	63	Wh/kg
Energy Density	131	70.9	117	Wh/L

8

Chemistry Consumption For Storage

Market Will Exceed 1 TWh by 2024

Chemistry	# of Batteries	Nickel Zinc		Zinc Air
		Zinc Oxide	Nickel Hydroxide	Zinc Oxide
Lbs. Chemistry/Battery	1	10.2	11.6	tbd
Chemistry per MWh	500	5,119 lbs.	5,802 lbs.	5,168 lbs.
Chemistry per GWh	500,000	5,119,000 lbs.	5,802,000 lbs.	5,168,000 lbs.
Metric Tons / GWh		2,327	2,637	2,349
Metric Tons / TWh	500,000,000	2,327,000	2,637,000	2,349,000
% of Global Production		136% of ZnO 9% of Zinc Metal	??% of Ni(OH)2 60% of Ni Metal	136% of ZnO 9% of Zinc Metal

Markets

Lead-Acid industries which Nickel Zinc can replace

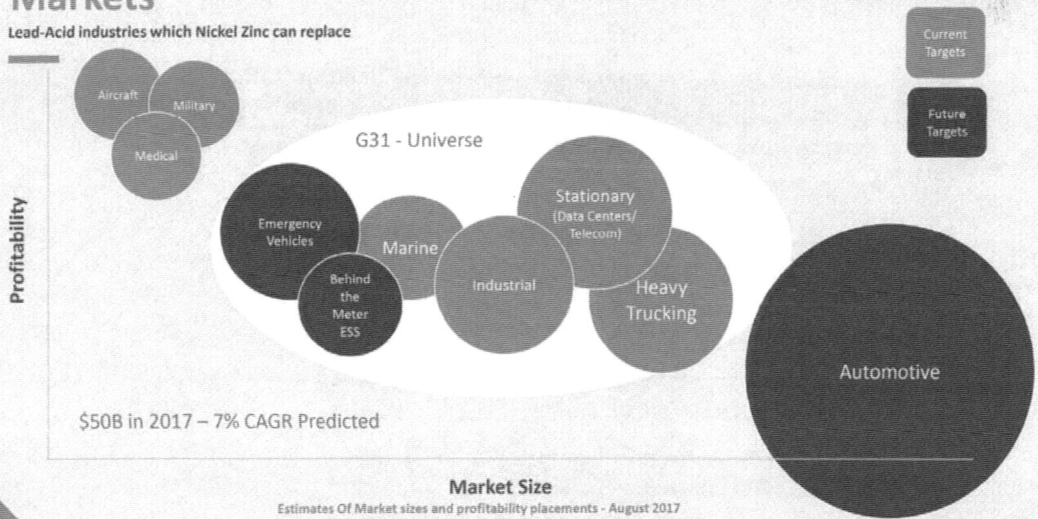

Estimates Of Market sizes and profitability placements - August 2017

10

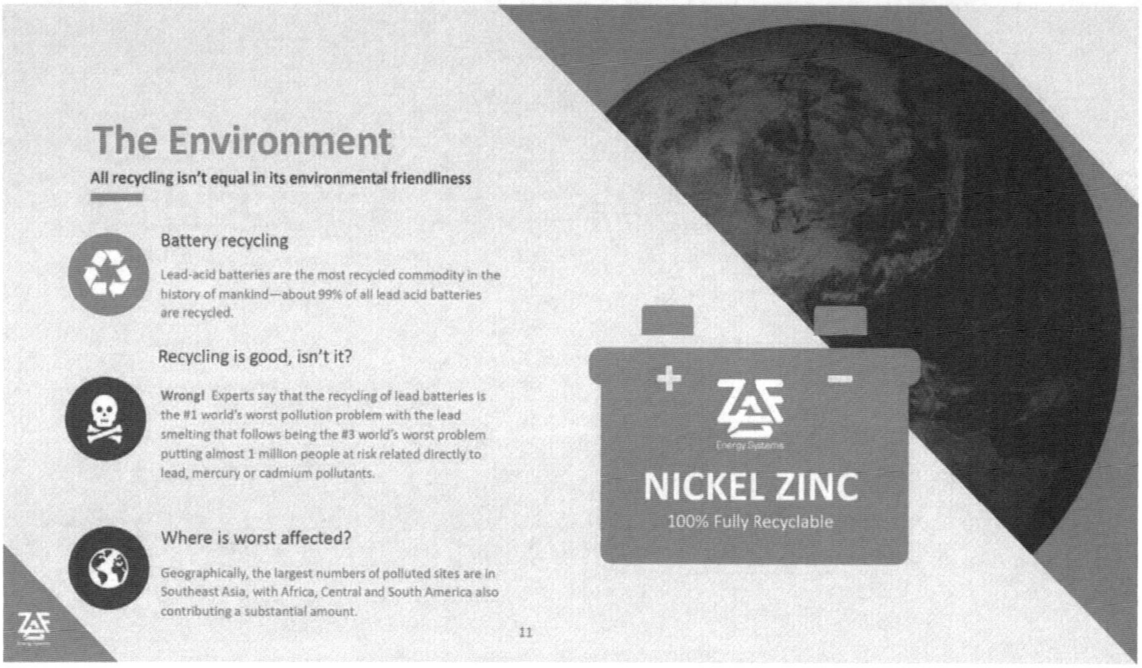
The Environment
All recycling isn't equal in its environmental friendliness
Battery recycling
Lead-acid batteries are the most recycled commodity in the history of mankind—about 99% of all lead acid batteries are recycled.
Recycling is good, isn't it?
Wrong! Experts say that the recycling of lead batteries is the #1 world's worst pollution problem with the lead smelting that follows being the #3 world's worst problem putting almost 1 million people at risk related directly to lead, mercury or cadmium pollutants.
Where is worst affected?
Geographically, the largest numbers of polluted sites are in Southeast Asia, with Africa, Central and South America also contributing a substantial amount.
+
-
NICKEL ZINC
100% Fully Recyclable
11

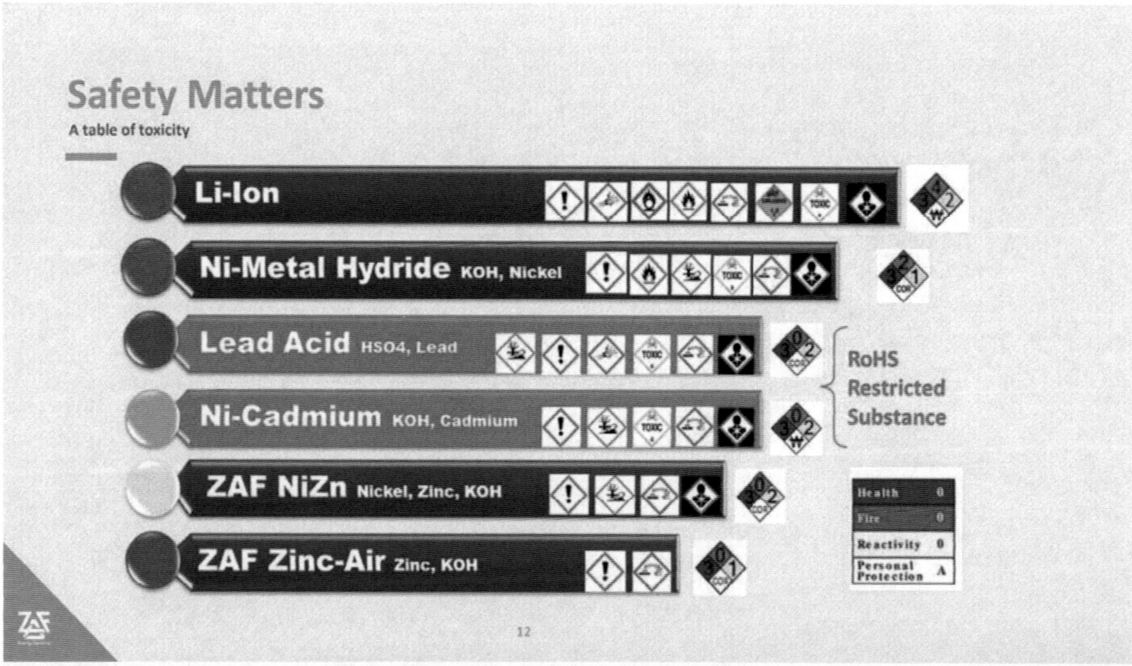
Safety Matters
A table of toxicity
Li-Ion
Ni-Metal Hydride KOH, Nickel
Lead Acid HSO4, Lead
Ni-Cadmium KOH, Cadmium
ZAF NiZn Nickel, Zinc, KOH
ZAF Zinc-Air Zinc, KOH
RoHS Restricted Substance
Health 0
Fire 0
Reactivity 0
Personal Protection A
12

Æsir Technologies, Inc.
Æsir Nickel-Zinc Aircraft Batteries
"Mission critical power for the most demanding applications™"
Case Study in NiZn Battery Application

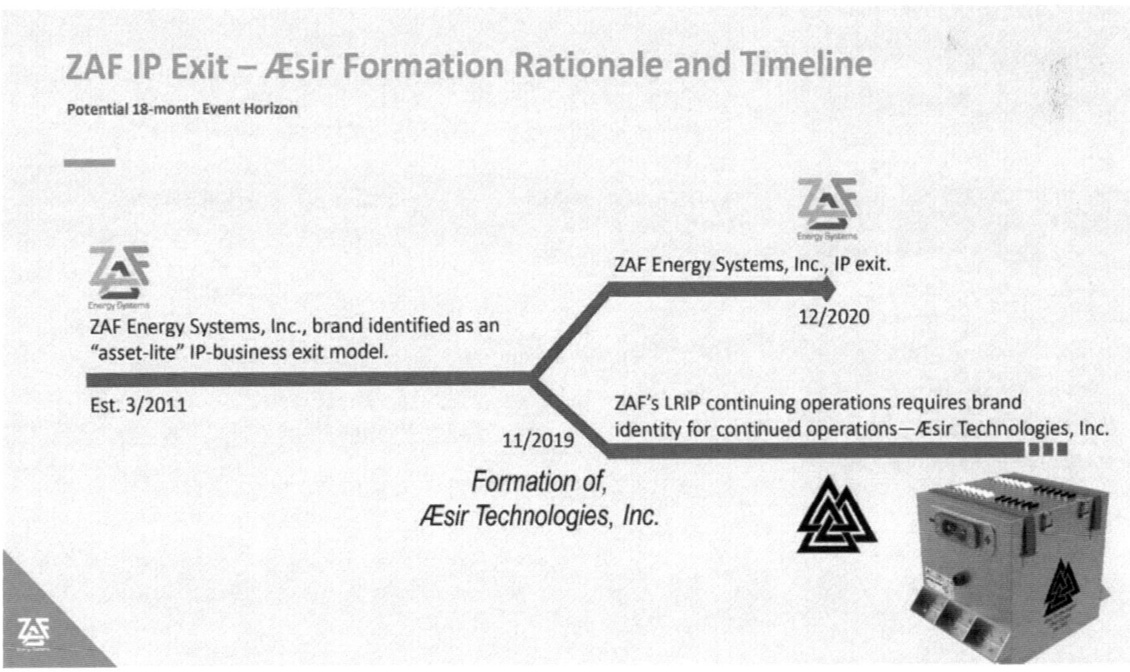
ZAF IP Exit – Æsir Formation Rationale and Timeline
Potential 18-month Event Horizon
ZAF Energy Systems, Inc., brand identified as an "asset-lite" IP-business exit model.
Est. 3/2011
ZAF Energy Systems, Inc., IP exit.
12/2020
11/2019
ZAF's LRIP continuing operations requires brand identity for continued operations—Æsir Technologies, Inc.
Formation of, Æsir Technologies, Inc.

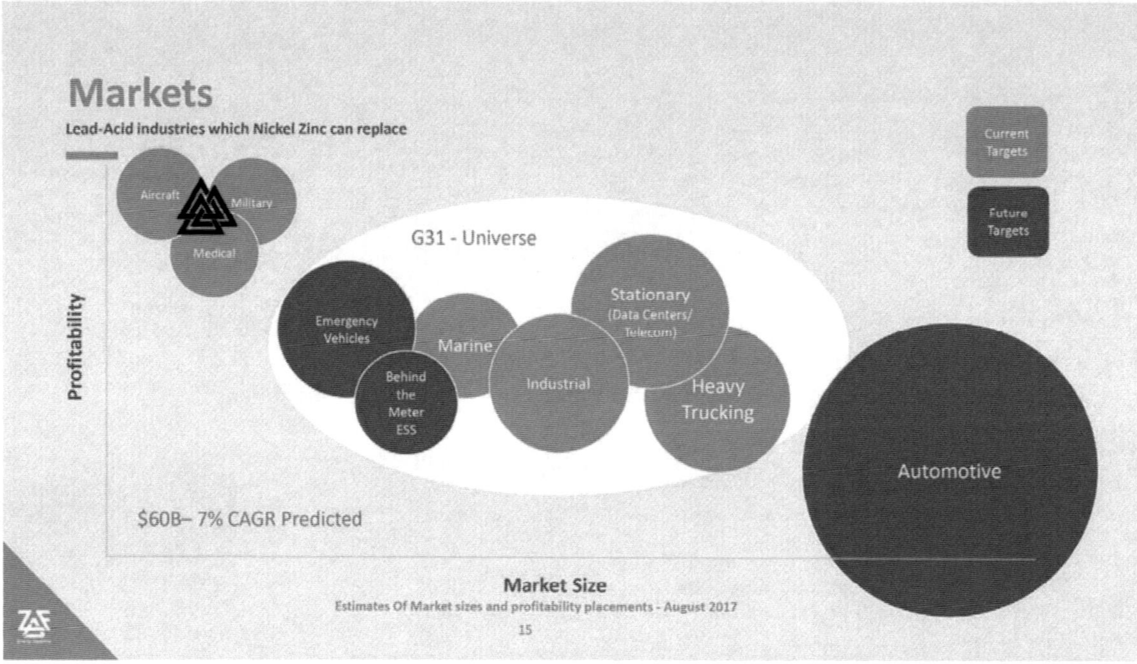
Markets
Lead-Acid industries which Nickel Zinc can replace
Current Targets
Future Targets
Aircraft
Military
Medical
G31 - Universe
Emergency Vehicles
Marine
Behind the Meter ESS
Industrial
Stationary (Data Centers/ Telecom)
Heavy Trucking
Automotive
Profitability
$60B– 7% CAGR Predicted
Market Size
Estimates Of Market sizes and profitability placements - August 2017
15

Market Opportunity-Aircraft Batteries
$400M Annual Market with Few Competitors and Ready for Disruption
• General Aviation (GA) Market Forces – NextGen avionics in light aircraft driving need for more power and energy at less weight.
• Business Aviation Market – Less price sensitive than commercial or GA. NiZn provides the performance improvements and reduced maintenance cost.
• Helicopter Market – PbA weight considerations have always been design problems. New chemistries provide opportunities to increase useful payloads.
• Commercial Aircraft Market – Benefits of Lithium Ion over NiCd driving migration, but "safe" and capacity enhanced solutions provide opportunity for nickel-zinc.
Past and Future Product Solutions
Commercial
Business
Helicopters
General Aviation
Past
Present
Future
Ni
Cd
Li
Fe
Pb
ACID
Ni
Zn
10

Why Nickel Zinc Batteries?

Why Æsir?

Æsir's nickel-zinc battery design solves historic NiCd problems.

- ✓ **Memory effect -> inherently solved because the NiZn combination does not suffer the effect**
- ✓ **Electrolyte dry out -> solved through recombination device**
- ✓ **Environmental issues -> NiZn contains no cadmium, a heavy metal increasingly difficult to recycle and subject to governmental regulations banning hazardous substances**

Battery Data	NiCd	ZAF NiZn Capacity G31	Unit
Memory Effect	Yes	No	
Self Discharge	20%	5-7%	Per Month
100% DOD Life Cycles	500	600+	cycles
60% DOD Life Cycles	1,100	1,900	A
specific energy	40-60	87	Wh/kg
specific power	150	587	W/kg
energy density	50-120	180	Wh/L
power density	800-1000	1161	W/L

Why Nickel Zinc Batteries?

Why Æsir?

Æsir's nickel-zinc battery design solves emerging Li-ion problems.

- ✓ **Cost -> NiZn is considerably less expensive than Li-ion (no BMS, no DO-178, no DO-311)**
- ✓ **Safety -> NiZn is non-flammable**
- ✓ **Weight -> Nizn is lighter than FAA DO-311 compliant Li-ion**
- ✓ **Environmental issues -> NiZn is economically recyclable**

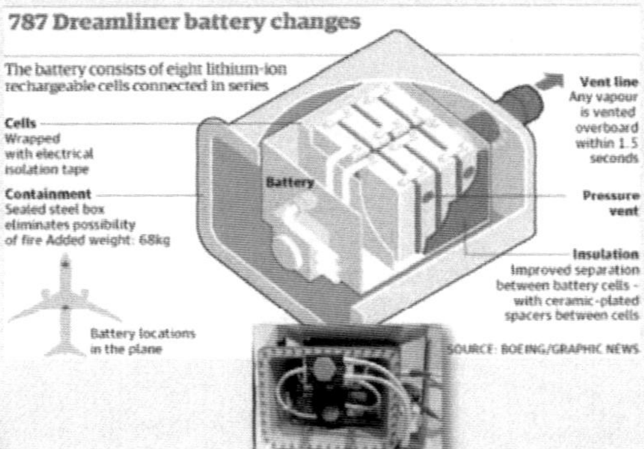

18

Naval

Underwater power

Naval underwater batteries are a $50M+/yr market

- **Submarines have on-going battery issues**
  - Formerly flooded lead acid batteries lasted 7 years
  - Now VRLA lead acid batteries last 3-4 years
  - New Columbia-Class submarine's systems too much load for lead acid
- **NAVSEA worked with major lead-acid company**
  - Using competitor's NiZn technology, reportedly not getting the job done.
  - NAVSEA approached ZAF, Development Contract now in place.

NICKEL ZINC

ZAF currently meets or exceeds current specifications

19

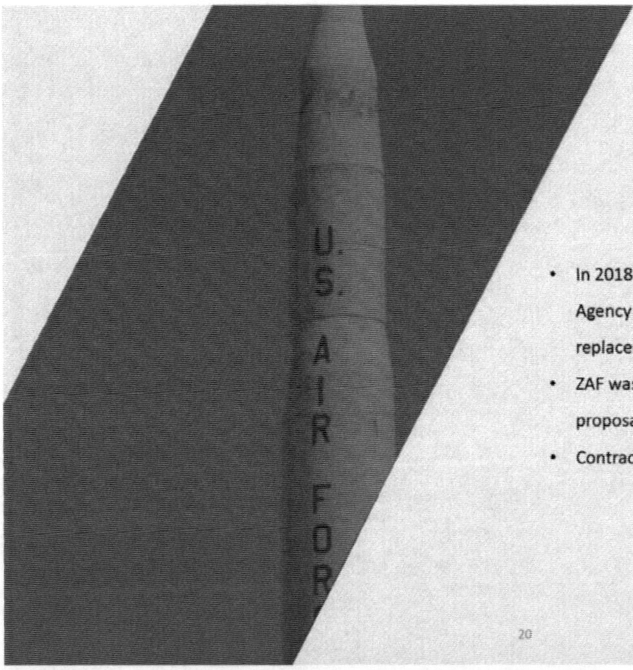

Contract Award

Minuteman III, ICBM Silo Ground-Station Backup Power

- In 2018 ZAF Proposed to the Rapid Innovation Fund (RIF) Broad Agency Announcement (BAA) titled "ICBM battery: ZAF Ni-Zn replacement batteries for ICBM ground facility battery".
- ZAF was notified by the Air Force proposal review team that our proposal had been selected for contract award.
- Contract Value: $1.4M

U.S. AIR FORCE

20

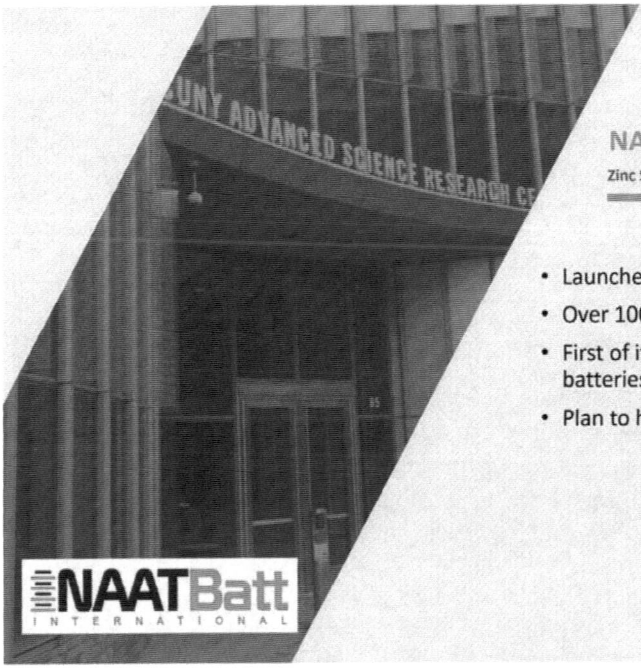

NAATBatt

Zinc Subcommittee Successes

- Launched Zinc Battery Workshop in 2018
- Over 100 attendees at each annual event
- First of its kind organization focused on zinc-based batteries
- Plan to have third workshop in Joplin, MO, Q4 2020

ZAF Energy Systems

Summary

- We have the ideal technology and a compelling value proposition to displace lead-acid.
- We address a huge and growing market.
- We have a formula for sustainable energy storage unlike any other competing product.
- We have the strategic partnerships and business strategy to scale production to meet demand.
- We have the plan and the team to execute.

Energy Systems

22

World Economic Forum Technology Pioneer 2019

OPENMINERAL

Open Mineral pioneering technology

Industry relationships, market intelligence and technology allows us to create a unique commercial and market intelligence platform

1. **Digital platform for sellers and buyers** – OM sources deals, takes title and accumulates market intelligence
2. **Large evolving database of product qualities as well as sellers & buyers with their specific preferences**
3. **Proprietary pricing algorithms** – OM estimates the value of each concentrate on current market terms
4. **Real-time market intelligence** – OM provides valuable content: unique industry research and analytics as well as price indices
5. **Efficient, carefully targeted communication tool** – OM proactively reaches out to companies with offers to buy / sell products, playing a market maker role
6. **Unique opportunities empowered through data analytics** – OM identifies opportunities through value adding combinations of products

Physical trade of metal concentrates and secondaries

powered by technology
and market intelligence

2

Long-run TC/price relationship breaks and TCs go too far?

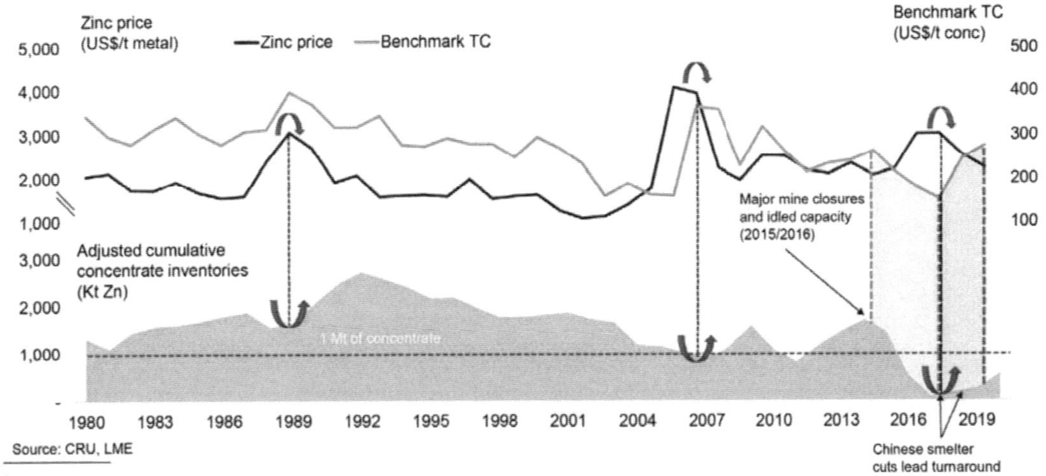

Source: CRU, LME

Market expectation has driven TCs and price

1. Expectations of mine supply growth in 2018 H2, 2019 were far higher than reality.
2. Concentrate balances since turnaround in 2018 Q2/Q3 suggest TCs over reacted.
3. Metal prices have possibly taken stronger cue from concentrate markets – Why? Iceberg effects? And expectation of natural flow through from conc-metal.

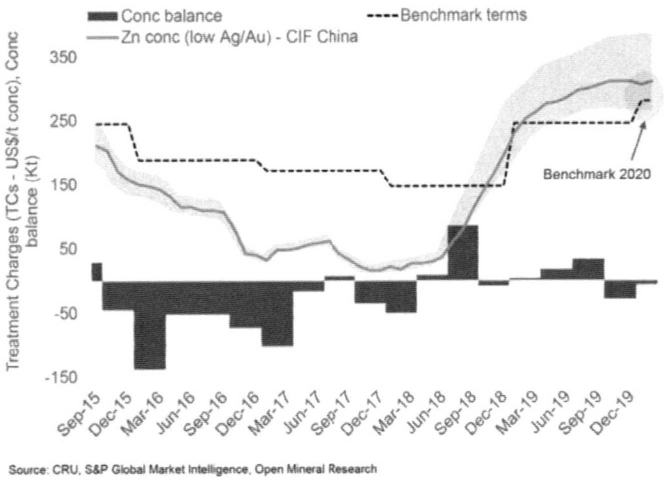

Source: CRU, S&P Global Market Intelligence, Open Mineral Research

World Economic Forum Technology Pioneer 2019

Discussion points

Some macro trends and demand

Key drivers for prices and TCs

Conclusions

5

World Economic Forum Technology Pioneer 2019

US bond market inversions and recessions

1. Historically, when the bond yields of 2- & 10-year treasury bonds have crossed, a recession has followed.
2. Inversion occurred in August 2019.
3. Base metals demand growth continued to slow through 2018/2019.

Source: Federal Reserve Bank of St Louis

6

World Economic Forum Technology Pioneer 2019

US bond market yield curve and slowing growth

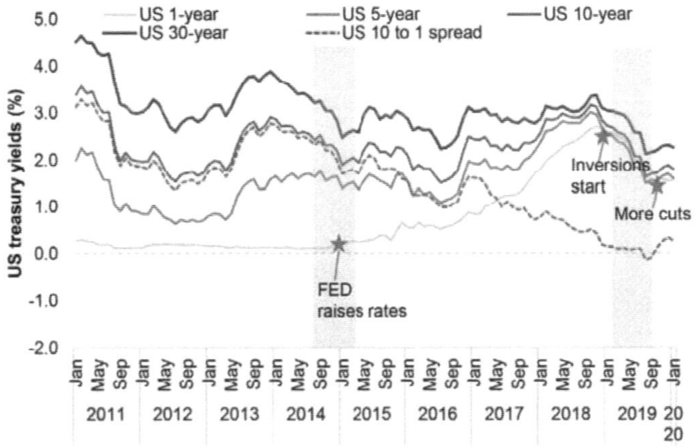

Source: Federal Reserve Bank of St Louis

7

Federal Reserve cuts rates again amid trade and growth fears

30 October 2019

The Fed's trying to interpret mixed economic signals

The US central bank has cut interest rates again, hoping to shield the economy from the impact of trade wars and a global slowdown.

The Federal Reserve lowered the target for its benchmark rate by a quarter point, to a range of 1.5% to 1.75%. The move was the third cut in four months

World Economic Forum Technology Pioneer 2019

German bond market and negative yields

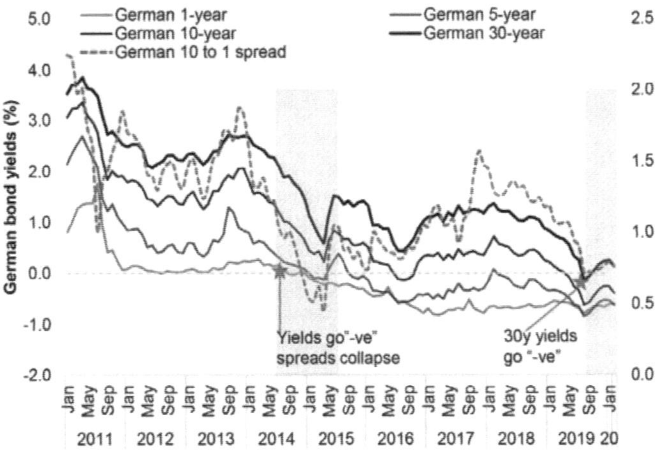

Source: Federal Reserve Bank of St Louis

8

THE WALL STREET JOURNAL

Germany sells nearly $1 billion of 30-year negative yielding bonds

Published Aug 21, 2019 7:29 a.m. ET

By PAUL J. DAVIES

Olaf Scholz, Germany's finance minister

Donald J. Trump @realDonaldTrump

So Germany is paying Zero interest and is actually being paid to borrow money, while the U.S., a far stronger and more important credit, is paying interest and just stopped (I hope!) Quantitative Tightening. Strongest Dollar in History, very tough on exports. No Inflation!....

93.4K 1:56 PM - Aug 21, 2019

30K people are talking about this

Chinese coronavirus – eroding demand, driving surpluses

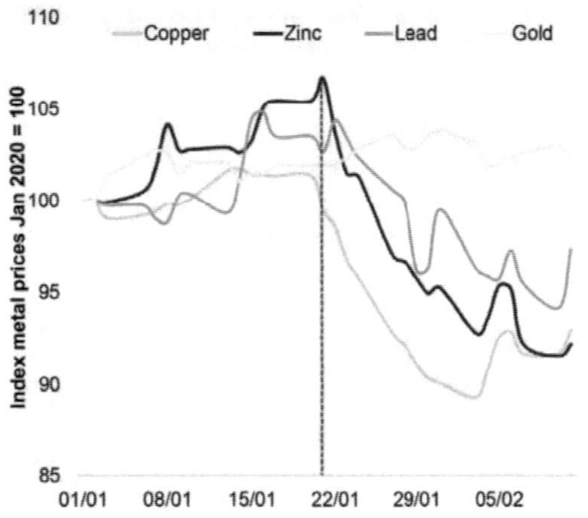

Source: LME, LBMA, news articles - Fastmarkets, ING, Antaike

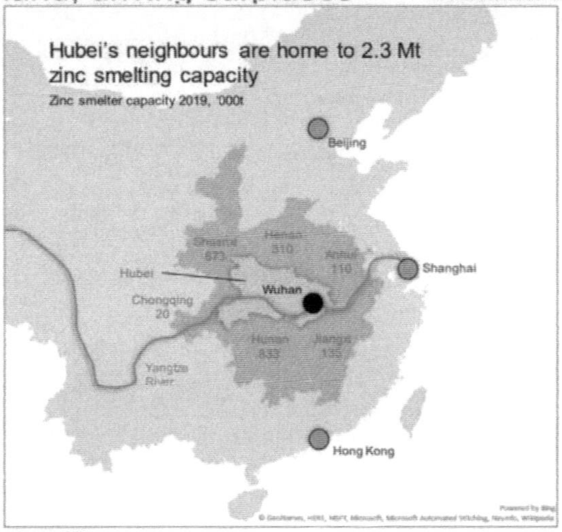

Source: CRU

9

Flatter yield curves, plummeting PMIs and two consecutive years of contraction in zinc consumption

Source: Trading Economics

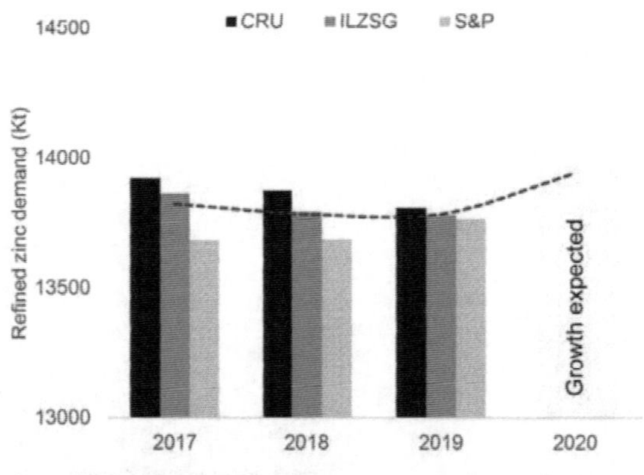

Source: CRU, ILZSG, S&P Global Market Intelligence

10

World Economic Forum Technology Pioneer 2019

Discussion points

- Some macro trends and demand
- Key drivers for prices and TCs
- Conclusions

11

World Economic Forum Technology Pioneer 2019

Zinc supply – Mine cuts could dampen growth expectations

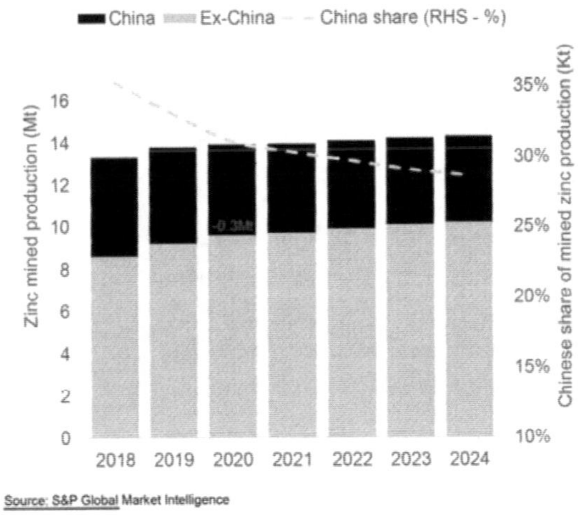

Source: S&P Global Market Intelligence

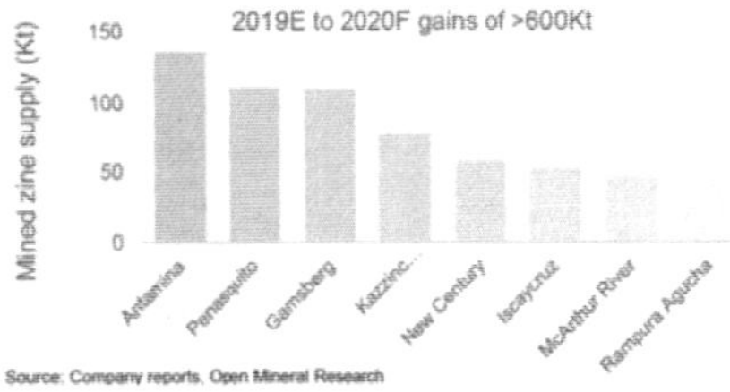

Source: Company reports, Open Mineral Research

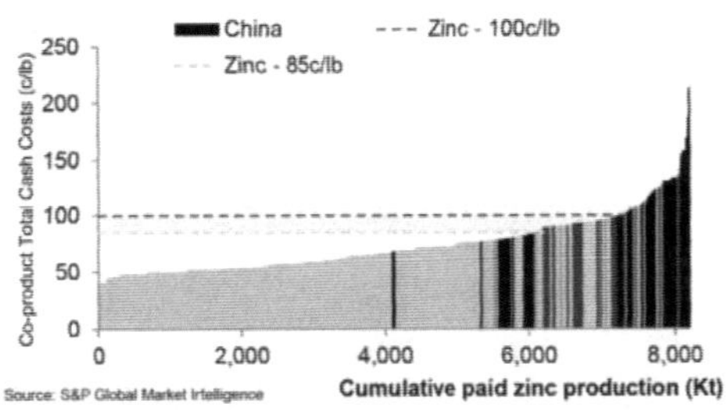

Source: S&P Global Market Intelligence

12

World Economic Forum Technology Pioneer 2019

Declining Chinese mine supply to drive concentrate imports

OPENMINERAL

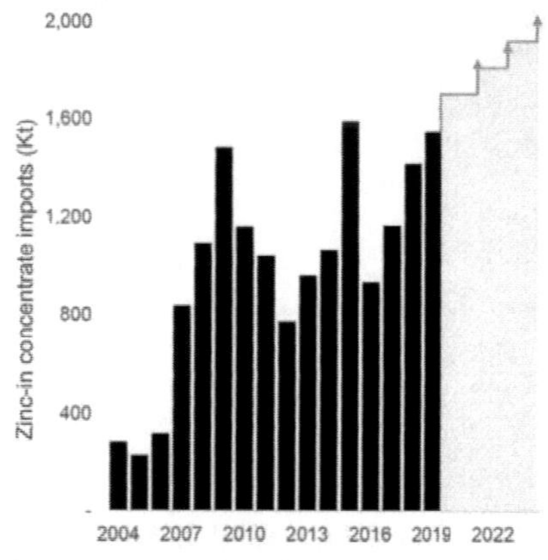

Source: CRU (historical data), Open Mineral Research

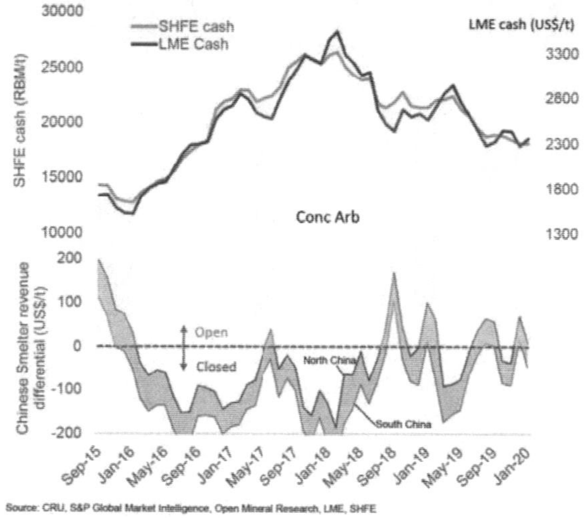

Source: CRU, S&P Global Market Intelligence, Open Mineral Research, LME, SHFE

13

World Economic Forum Technology Pioneer 2019

TCs – how much further can they run up?

OPENMINERAL

What we know

1. Expectations of mine supply growth in 2018 H2, 2019 were far higher than reality. Coordinated smelter cuts turned market and surpluses are modest compared to TC levels.
2. Chinese smelter revenue share ~40% in August 2019, and currently at high 30s, although outurn domestic TCs have been fallen through 2019 H2.
3. CIF China Spot TCs currently >>$315/t conc for standard material.

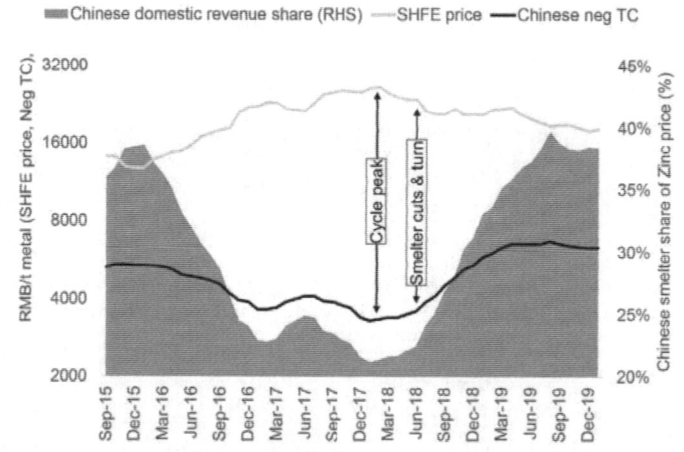

Source: CRU, SHFE, Open Mineral Research – *assumes revenue share based on 100% domestic feed*

14

Visible stocks still low, but expectations are shifting

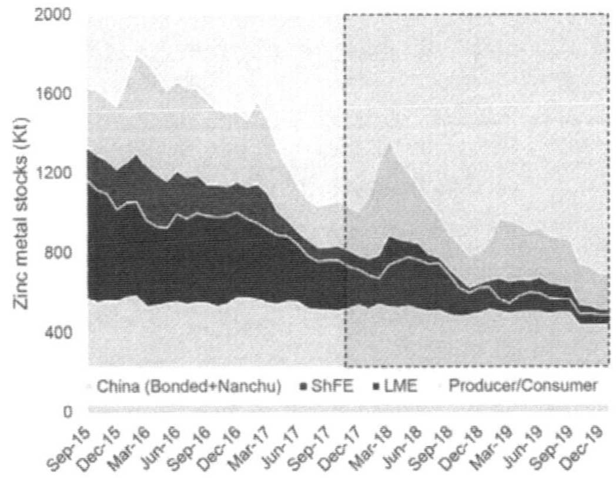

Source: CRU, S&P Global Market Intelligence, ILZSG

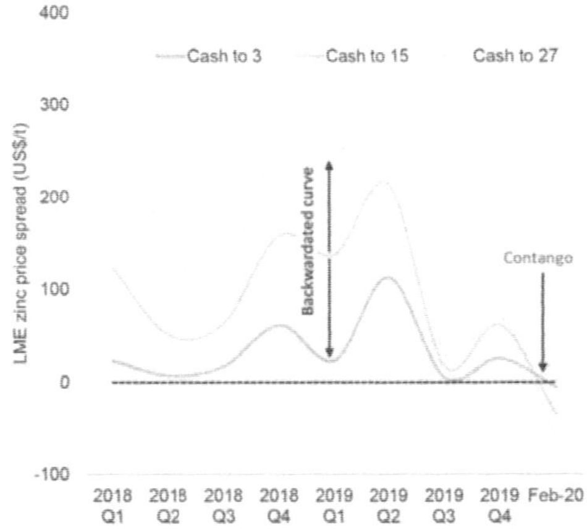

Source: CRU, S&P Global Market Intelligence, LME

15

Smelter profitability remains strong and refined deficits appear a thing of the past

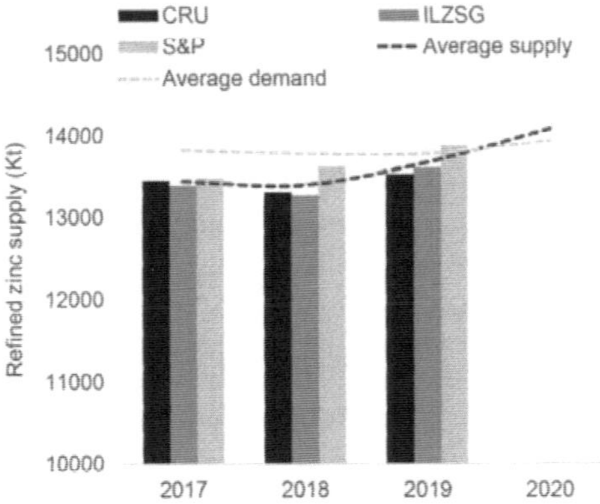

Source: CRU, ILZSG, S&P Global Market Intelligence

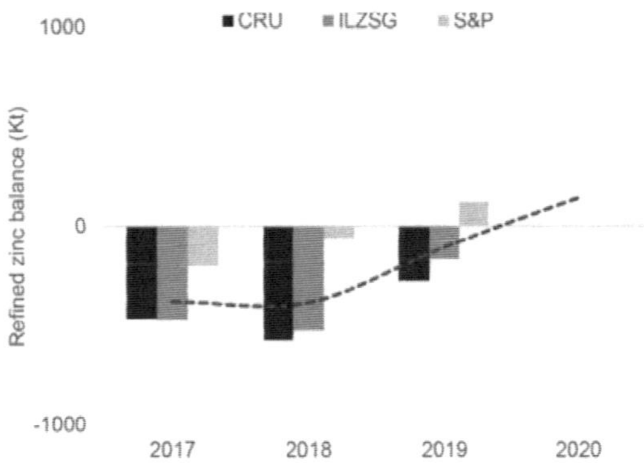

Source: CRU, ILZSG, S&P Global Market Intelligence

16

World Economic Forum Technology Pioneer 2019

But…german-US yield spreads suggest TWI to weaken

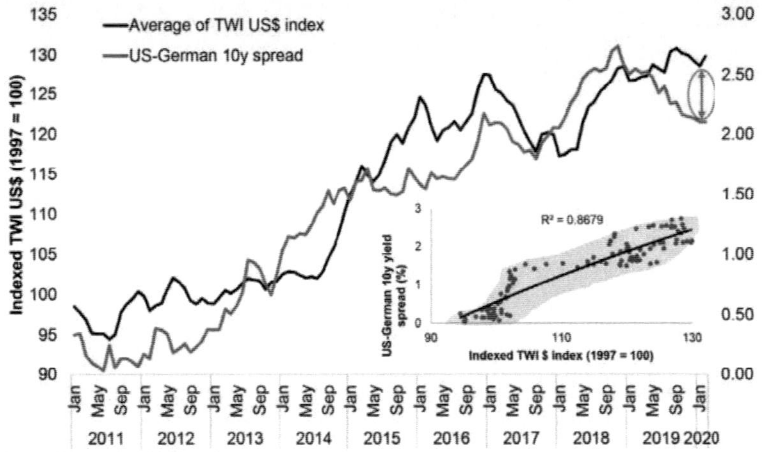

Source: Federal Reserve Bank of St Louis

1. High degree of correlation seen between TWI and US-German yields.
2. The variation in spreads has arguably been a leading indicator for TWI.
3. Recent uptick in the USD followed coronavirus outbreak

17

World Economic Forum Technology Pioneer 2019

And base metals still priced lower than 2011 while Dollar is 30% stronger

1. Copper exhibits the strongest of the inverse correlations to TWI $ index.
2. Pre-coronavirus expectation was broadly for weaker dollar through 2020 as a first phase of trade deal between US and China was signed.

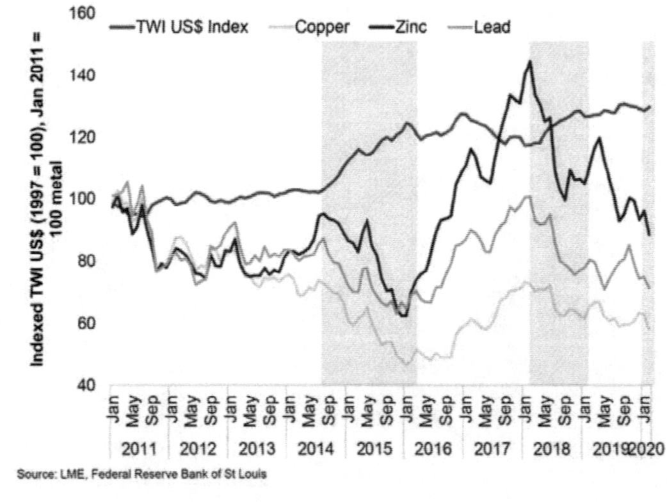

Source: LME, Federal Reserve Bank of St Louis

18

Price expectations - challenging year ahead for zinc miners

OPENMINERAL

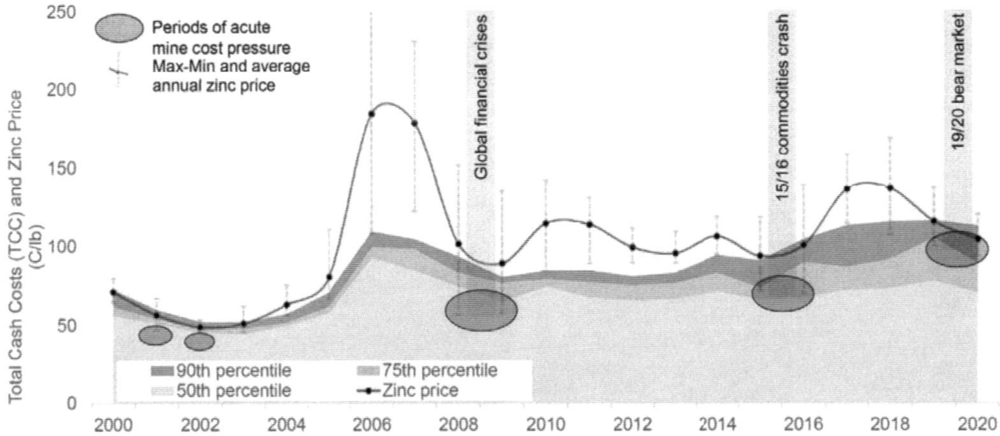

Source: S&P Global Market Intelligence, LME, *Notes - Total Cash Costs - co-product basis expressed in real 2019$ on a paid metal basis*

Conclusions – how much further can TCs run up

OPENMINERAL

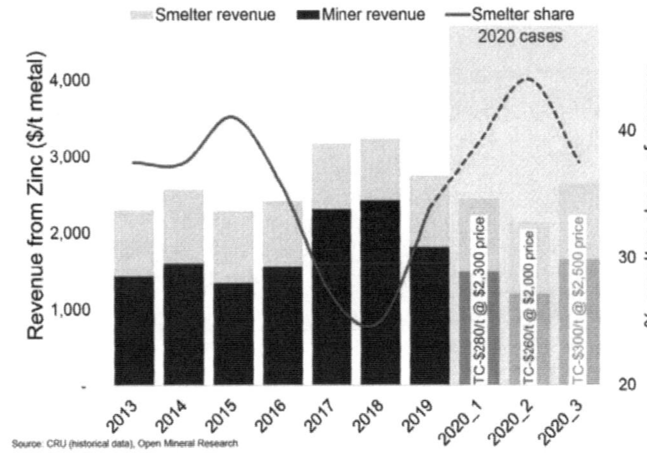

Source: CRU (historical data), Open Mineral Research

Backdrop for 2020

1. Coronavirus is eroding metal demand is likely to support high spot TCs, lower Chinese smelter utilization rates for 2020 Q1.
2. Expectations for 2020 mine supply growth considerably outstrip demand expectations – ***again***. NCT, Gamsberg, Antamina, Penasquito key incremental suppliers for 2020. **Cuts will be a key offsetting theme.**
3. Visible refined metal stocks still at decade lows and once… coronavirus effects begin to abate; USD could be poised to weaken supporting commodities including zinc.
4. All except the direst scenarios suggest upside to current spot price levels and suggest Ex-Chinese TCs could settle around ~$280/t conc @2,300 price.

World Economic Forum Technology Pioneer 2019

OPENMINERAL

Our team

Boris Eykher
CEO & Co-Founder
GLENCORE
HARVARD BUSINESS SCHOOL

Ilya Chernilovskiy
COO & Co-Founder
GLENCORE
HARVARD BUSINESS SCHOOL

Marc Pfefferli
Financial Advisor / CFO
GLENCORE
MET
TRAFIGURA

Ryan Cochrane, PhD
Head of Research
CRU
Wood Mackenzie

Diego Molina
Business Development, South America
GLENCORE
LEHIGH

Jonathan Fruchter
Business Development, North America
Gerald Group

Edouard Penot
Business Development, Europe and Australia
Morgan Stanley

Wen Wang
Business Development, China
GLENCORE
MRI

Roger Steiner
Head of Contract Execution and Risk
GLENCORE

David Gomez
Operations South America
GLENCORE
HOCHSCHILD MINING

Nikita Yampolski
Head of Product
gemalto
Yandex

Olga Serdiukova
Head of Marketing
Microsoft

World Economic Forum Technology Pioneer 2019

OPENMINERAL

Open Mineral at a glance

40%	Companies registered on the platform represent over 40% of metal concentrate industry globally
400+	Open Mineral registered members: 400+ metals & mining companies as well as traders from 39 countries
4000+	Gained industry credibility. 4000+ industry professionals. Companies reach out with opportunities.
#DISRUPTMINING	Recognized as "the" industry innovation. Finalist of 2018 #Disrupt Mining competition
WORLD ECONOMIC FORUM	In addition to sourcing, built the full capability to execute sourced contracts for clients - end-to-end solution Awarded 2019 Technology Pioneer by World Economic Forum

MOLYMET. GOLDCORP Sumitomo Corporation RioTinto nyrstar POLYMETAL umicore UMMC VOLCAN HOCHSCHILD MINING FORTUNA BBH SUMMIT LINCUNA NORNICKEL ZIJIN Numon FRESNILLO XJ

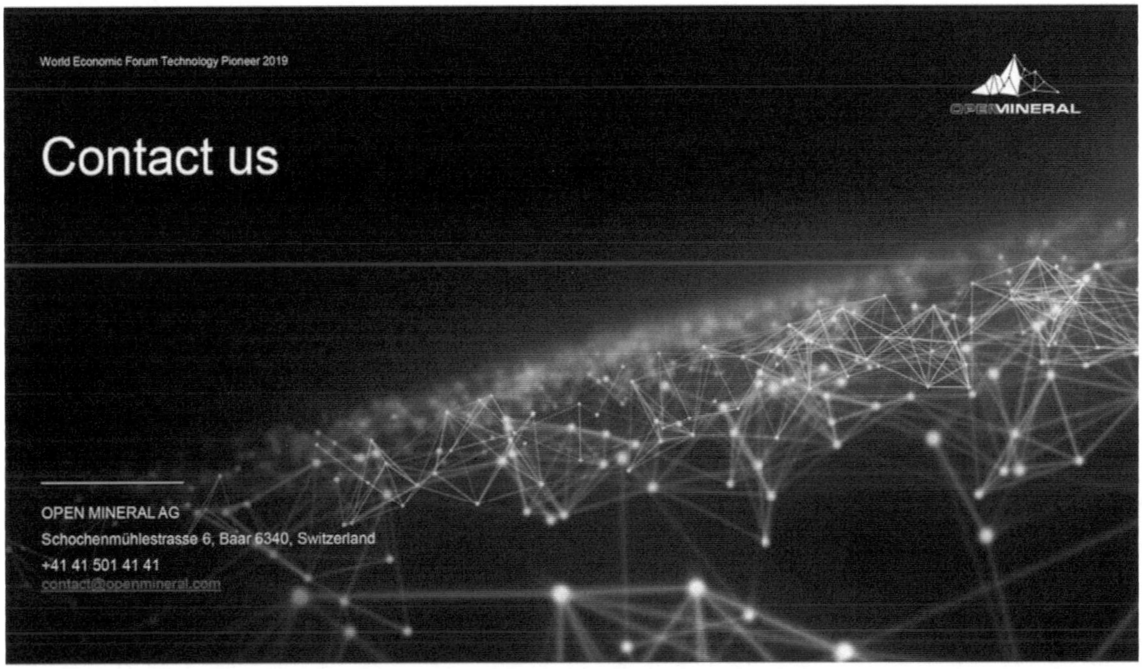
World Economic Forum Technology Pioneer 2019
OPENMINERAL
Contact us
OPEN MINERAL AG
Schochenmühlestrasse 6, Baar 6340, Switzerland
+41 41 501 41 41
contact@openmineral.com

Regulatory Update: Challenges to the Market

Eric Van Genderen & Christine Spirlet
International Zinc Association

Chemical Management

- Global
  - GHS, IMO, Hg, Sunscreen,
- Americas
  - "CEPA toxic", GHS, tires
- Asia-Pacific/ROA very active
- Europe
  - REACH, Pb, VMP, RoHS, Brexit, ED, Green Deal

2

Global Chemical Management

IMO issues

INTERNATIONAL MARITIME ORGANIZATION

- **Cargo corrosivity**: support for proposed changes to Chapter 9 of the IMSBC and the MHB(CR) test protocol.
- **Self heating** properties of concentrates are discussed
- New schedule for **lead concentrate** proposed by Belgium
- Proposal to amend Section 3 of the IMSBC Code with respect to **risk of entering confined spaces.**
  - Since 2010, 140 workers died in related incidents!
- MEPC:
  - Low Sulphur fuel from 1/2020: new risks? Increased shipping cost
  - Polar Code: emission could affect use of polar route.

Reports from COP3 (November 2019)

- Mercury Waste Expert Group
  - ➢ Technical analysis of threshold options
  - ➢ Determine thresholds for tailings from non-ferrous metal mining
  - ➢ Revision of list of mercury waste
- Scope for Releases Expert Group
  - ➢ Not include point sources addressed in other provisions of the Convention
  - ➢ Diffuse sources should not be included
  - ➢ Management of waste address releases to land and water

Sunscreen Bans Benefit Zinc

Another Tropical
Sunscreen Ban

Coral: Palau to ban sunscreen products to protect reefs

'We have one
sunsc

Originally publish

Lawmakers in US Virgin Islands approve bill banning certain sunscreen

Aruba Officially Bans Plastics and Oxybenzone

By Marcella Janczewski
November 21, 2019
Eco-friendly Aruba, General, The Aruban Community

November 19, 2019 marked a very important day for Aruba's future! The Parliament of Aruba has officially banned single use plastics and Oxybenzone, a chemical found in many sunscreens that contribute to the deterioration of coral reefs.

n in July.

North America

Canada

- Draft CMP3 Zn Assessment – "CEPA toxic"

USA

- Draft MSGP update released last week!

California

- DTSC "Zinc in Tires" petition ongoing
- Prop 65 Safe Use Designation for all metallic zinc applications

CMP Risk Management

- Proposed Schedule 1 designation
- Risks of concern based on metal mining and smelting/refining
- Preventative and control actions
  - ➢ As appropriate and economically feasible
- IZA and stakeholder response highlighted:
  - ➢ Errors in screening assessment
  - ➢ Bioavailability not appropriately considered
  - ➢ Existing instruments cover emissions of concern

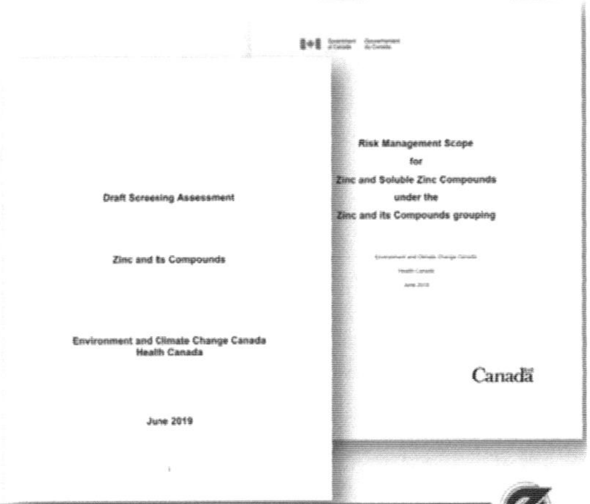

Draft Screening Assessment

Zinc and Its Compounds

Environment and Climate Change Canada
Health Canada

June 2019

Risk Management Scope
for
Zinc and Soluble Zinc Compounds
under the
Zinc and its Compounds grouping

Canada

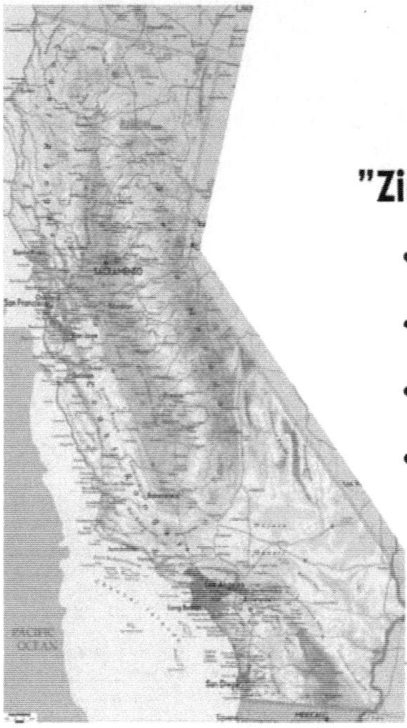

California – DTSC

"Zinc in Tires"

- May 2018 – CASQA filed petition
- July 2018 – IZA/USTMA arguments
- Jan 2019 – DTSC Supplemental Info Request
- Mar/Apr 2019 – IZA/USTMA advocacy

U.S. TIRE MANUFACTURERS ASSOCIATION CASQA Department of Toxic Substances Control

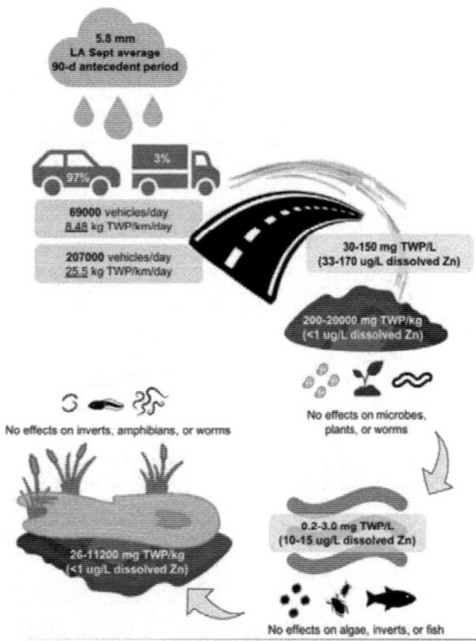

Full RA Submitted

- TWP fate pathway is quantifiable and predictable
- No effects of relevant direct TWP exposures
- Zinc from TWP does not exceed environmental quality standards

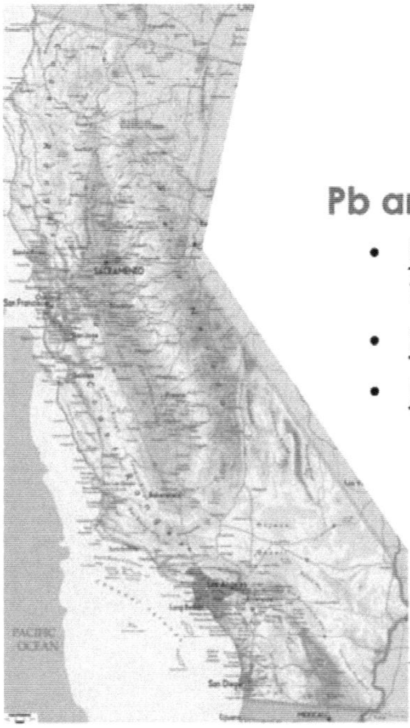

California – Prop 65

Pb and Cd exposures in consumer products

- Before: Metallic zinc impurities targeted as basis for financial penalties
- Now: "Position Paper" available
- Next: CA "Safe Use Designation" petition

Application	Maximum Pb (% by wt)	Maximum Cd (% by wt)
Galvanizing	0.003 – 1.4	0.003 – 0.2
Die Casting	0.003 – 0.006	0.002 – 0.006
Rolled Zinc	<0.005	<0.005

International Zinc Association
11

Latin America

MEXICO
- Implementation in 2018
- Comprehensive

COSTA RICA
- Implementation in 2017
- 5-year transition period

ECUADOR
- Implementation in 2013

COLOMBIA
- Most progressive
- Update of labels and SDSs every three years

CHILE
- Classification, Labeling, and Notification

GUATAMALA
HONDURAS
EL SALVADOR
- In process of adopting GHS

ANDEAN BLOC
MERCOSUR
- No consistency
- Food packaging

BRAZIL
- RoHS
- Chemical registry
- Transport regulations

ARGENTINA
- Workplace and waste
- Directorate for chemicals

International Zinc Association
12

Australia/NZ
- 7th GHS – July 2020
- AICIS replaces NICNAS

China
- ERA Guidance
- Risk Mgmt Guidance
- Native language limitation

Korea
- Dec 2021 first registrations
- Challenges for direct copy-paste

India
- National Action Plan
- Embargoed GHS plan?

Philippines
- Cadmium in focus?

Taiwan
- Registration for >10 tonnes
- ZnO and ZnCl2 listed

Thailand
- TECI (hazard-based)
- Authoritative prioritization

Vietnam
- NCI
- Priority nominations (2018/19)
- ASEAN database

Malaysia
Sierra Leone
UAE
- Chem Mgmt coming

13

REACH

ZnO evaluation

Germany is evaluating member state – aim to identify potential specific effects in ZnO nanomaterial

Major testing program launched both for environment and human health, including mutagenicity and toxicity to reproduction

Registration of nanoforms required as of January 1st 2020

MISA * (2020 update)

The approach provides a mechanism to improve REACH dossiers in formal collaboration with ECHA; Eurometaux coordinates effort

A substantial literature review and subsequent update of all substances is ongoing incl. exposure assessment and risk characterisation

Scrutiny of all registrations > 100T by 2023

International Zinc Association

* Metals and Inorganics Sectorial Approach

82
Pb
207.2

Pb metal classification for ENV

Pb classification (massive form) could have major impact due to cross-metal implications!

RAC opinion November 2018
- Classification of Pb metal ENV acute and chronic Cat 1
- No split between massive and powder forms
- Forwarded to COM for inclusion in ATP *

Opinion debated
- Based on non standard study (snail)
- Denied dissolution difference between forms
- New data provided by industry

Way forward
- COM will include only Pb powder in the 15th ATP
- COM asks RAC to review its opinion on Pb massive based on the new data

International Zinc Association

* Adaptation to technical progress

Endocrine disruption: emerging regulatory issue

Current fitness check of existing EU regulations (REACH, cosmetics, BPR, PPP, etc.)

Targeted stakeholders' consultation lead to the need of more data on reproduction toxicity, developmental neuro- and immunotoxicity

Study on Zn in the environment: effects (if any) occur at significantly higher levels than the PNEC* values

Plan to amend REACH annexes to reflect ED
- In the SDS (annex II), applicable by Dec 31st 2022
- In the data requirements by tonnage band (annexes VII to X) – proposal to come

International Zinc Association

* Predicted No Effect Level

ZnO ban in Vet meds – key facts

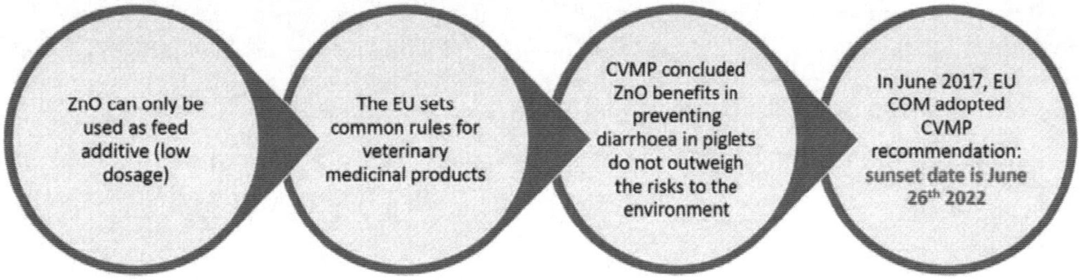

ZnO as feed additive (authorized)	ZnO as VMP (banned)
European Food Safety Authority (EFSA)	European Medicine Agency (EMA)
Regulation EC 1831/2003 on feed additives	VMP Directive 2001/82/EC and Regulation 726/2004
Max 150ppm total Zn	Normal dosage ca. 2500ppm

International Zinc Association

Is ZnO in RoHS list?

- ZnO is on a list which is an inventory of substances used/present in EEE* and (potentially) hazardous.
- Inventory (draft) constitutes a pool of substances from which the Commission could identify chemicals to be assessed for restriction and possible subsequent addition to Annex II of RoHS.
- Only when a chemical appears on Annex II is it effectively restricted (not banned).
- With the on-going 2021 RoHS Review, and the on-going assessment of seven shortlisted substances (which do not include ZnO), Industry does not expect that any (new) chemical will be shortlisted and proposed for restriction before 2022 at the very earliest.

International Zinc Association

*electric and electronic equipment

Brexit

Until Dec 2020, business as usual

Jan 29th 2020
Withdrawal Agreement Act

Feb 24th 2020
EU COM will receive its negotiating mandate from the Council

October 2020
Negotiations free trade agreement (FTA) between UK and EU COM expected to be completed

International Zinc Association

Brexit: myths and facts

Myth 1: All will change Feb 1st	**FALSE**, BAU through December 31, 2020 and UK will remain aligned with the EU and its regulatory regimes
Myth 2: FTA will insure frictionless trade between the UK/EU from 2021	**FALSE**, even if both sides wish to remain aligned on regulations, there will have to be customs (conformity with rules of origin) and delays and immigration will be challenging in some sectors
Myth 3:US clients need to do nothing	**FALSE,** US-UK trade negotiations will begin, in parallel with the UK-EU negotiations. Will affect both US clients trading into the UK and the EU, but also US clients that may be invested in the UK as a platform for servicing the EU
Myth 4: Everything will be in the FTA.	**FALSE**, there are expected to be a number of sectors that will not be included and these will be deferred until 2021

International Zinc Association

Commission's Green Deal and Zero-Pollution ambition

Climate neutrality by 2050

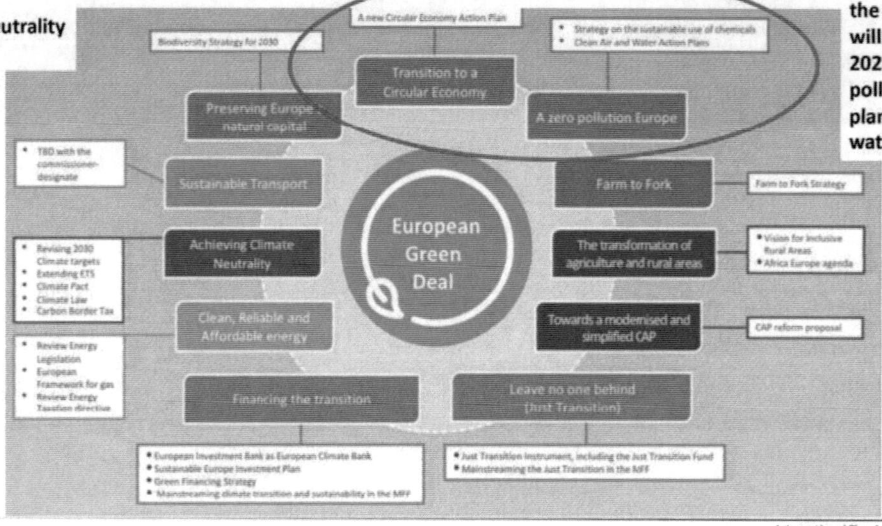

the Commission will adopt in 2021 a zero-pollution action plan for air, water and soil

International Zinc Association

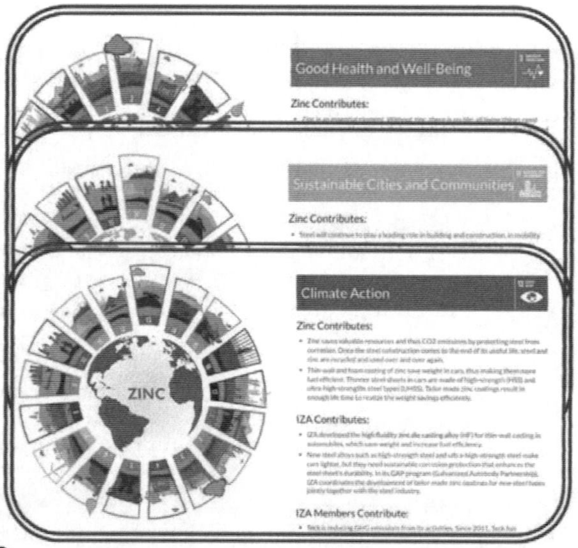

IZA Focus on UN SDGs

https://sustainability.zinc.org/

International Zinc Association

1

ARE YOU CONNECTED TO THE FUTURE?
MICHAEL TCHONG
UBERCOOL INNOVATION LLC

2

DESKTOP PUBLISHING INNOVATOR
INTERNET INNOVATOR
MacWEEK
CyberAtlas
ATELIER
ICONOCAST
ubercool INNOVATION
CRM INNOVATOR
DIGITAL MARKETING INNOVATOR
FUTURIST

3

" UBERTRENDS — HOW TRENDS AND INNOVATION ARE TRANSFORMING OUR FUTURE "
- MICHAEL TCHONG

This talk is all about u!
How trends and innovation will impact you in the 2020s
it's all about
U
ubercool.com

4

5

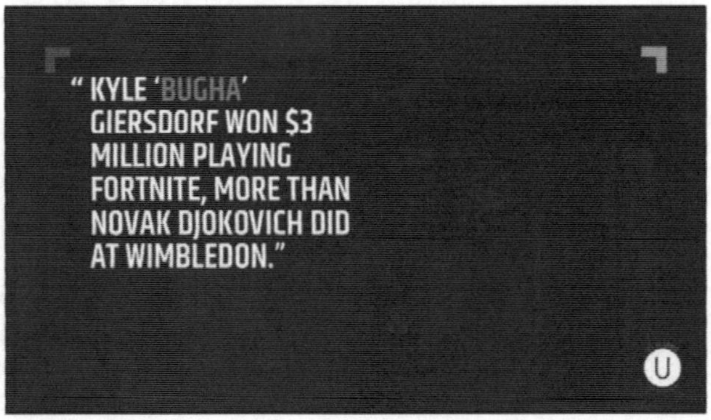
" KYLE 'BUGHA' GIERSDORF WON $3 MILLION PLAYING FORTNITE, MORE THAN NOVAK DJOKOVICH DID AT WIMBLEDON."

6

7

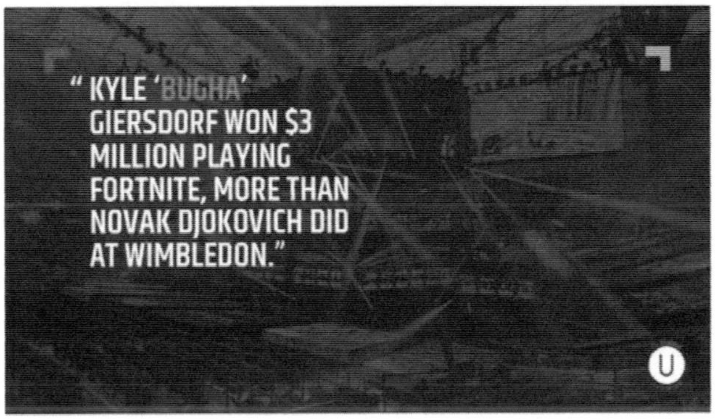
" KYLE 'BUGHA' GIERSDORF WON $3 MILLION PLAYING FORTNITE, MORE THAN NOVAK DJOKOVICH DID AT WIMBLEDON."

8

Boston Dynamics

9

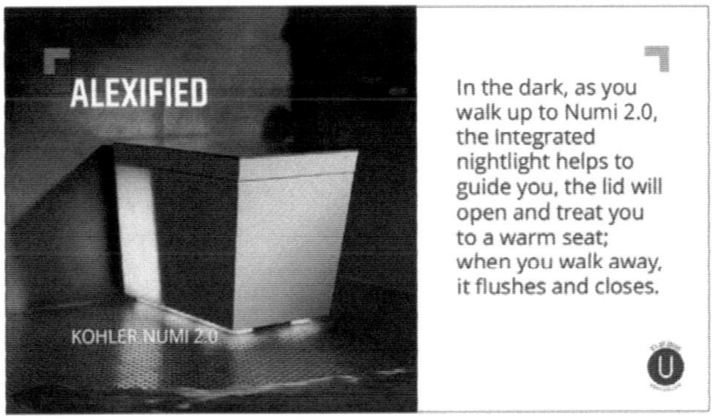
ALEXIFIED
KOHLER NUMI 2.0
In the dark, as you walk up to Numi 2.0, the integrated nightlight helps to guide you, the lid will open and treat you to a warm seat; when you walk away, it flushes and closes.

10

11

“Smart speakers’ global installed base is on track to top 200 million by the end of [2019].”

15-Apr-19 TechCrunch

12

LASER FOCUSED

13

"Two thirds of all trends are propelled by technology."

Michael Tchong, **Ubertrends**

14

15

Ubertrend

A massive wave, think tsunami, that cascades through society, leaving many subtrends in its wake.

16

What Sets an Ubertrend Apart?

An Ubertrend leads to **fundamental changes in societal values.** Most trends do not change values, they merely skim the social surface.

17

"A 16-year-old teen from Tampa, Fla. had **dropped her iPod into the path of an oncoming truck.** Thinking about it for a split second, she decided to save her iPod. Imagine her thought stream: "My **iPod or my life…iPod or life…iPod it is!**"

Michael Tchong, **Ubertrends**

18

Oxford's 2015 Word of the Year

"Face with Tears of Joy"

19

"David Levy is serious when he speculates that we'll be marrying robots by 2050 and that Massachusetts will be the first state to allow it."

Michael Tchong, **Ubertrends**

20

21

"Nearly half of Americans have cut back on spending, including for travel, food and health care, in order to afford their technology."

07-Oct-14 CNBC All America Survey

22

Spinali Design

Daisy Duke-style jeans send vibrations to different sides of body, indicating when and where to turn.

23

"Six-year-olds understand digital technology better than adults."

06-Aug-14 Ofcom, The Guardian

24

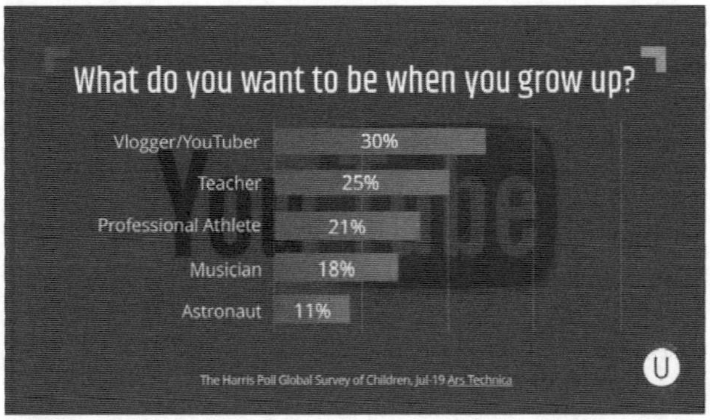

25

Thought Starter

The Digital Lifestyle is rewriting the rules of innovation. As technology becomes more tightly interwoven with the fabric of life, the dominant values ruling this tech-infused age are connectivity, convergence and convenience.

26

Digital Lifestyle Impact on Business

- Connectivity – All communication options should be *available online, preferably via chat and apps*
- Convergence – *Combine features*, i.e. download brochures, videos, etc., for optimum efficiency
- Convenience – Optimize the *user experience*

27

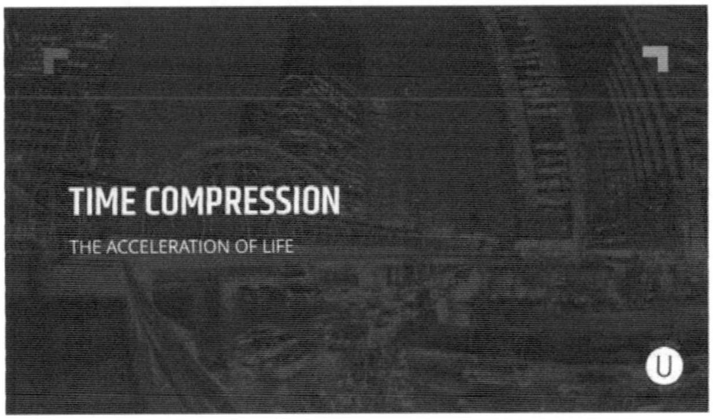

28

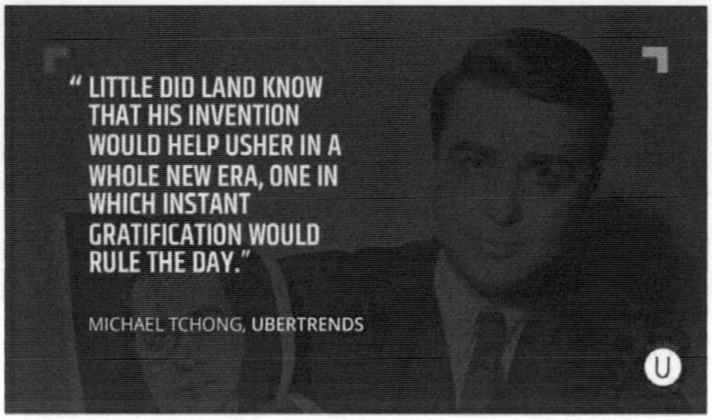
" LITTLE DID LAND KNOW THAT HIS INVENTION WOULD HELP USHER IN A WHOLE NEW ERA, ONE IN WHICH INSTANT GRATIFICATION WOULD RULE THE DAY."
MICHAEL TCHONG, UBERTRENDS
U

29

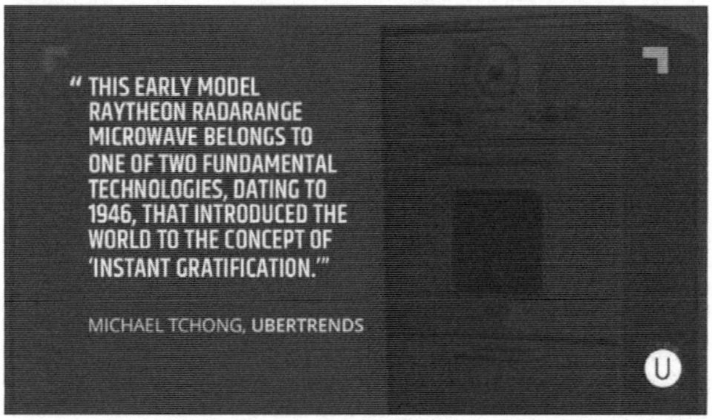
" THIS EARLY MODEL RAYTHEON RADARANGE MICROWAVE BELONGS TO ONE OF TWO FUNDAMENTAL TECHNOLOGIES, DATING TO 1946, THAT INTRODUCED THE WORLD TO THE CONCEPT OF 'INSTANT GRATIFICATION.'"
MICHAEL TCHONG, UBERTRENDS
U

30

TL;DR

75
seconds
10-May-13 "Duration of a Web page viewed," The Nielsen Company

31

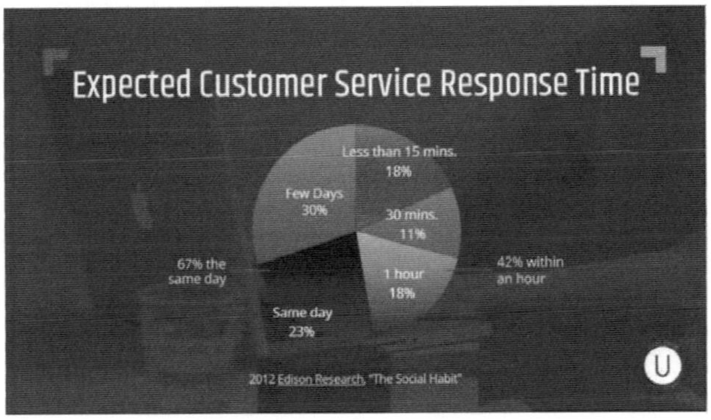
Expected Customer Service Response Time
Less than 15 mins. 18%
Few Days 30%
30 mins. 11%
67% the same day
1 hour 18%
42% within an hour
Same day 23%
2012 Edison Research, "The Social Habit"

32

"Google reports that searches for 'open now' have tripled in the past two years, while searches for 'store hours' dropped."
Michael Tchong, Ubertrends

33

34

Thought Starter

Time Compression has bred a multitasking population that's in love with instant gratification. It's a world that treasures speedy, efficient services that are simple to understand, and that better manage their increasingly valuable disposable time.

35

Time Compression Impact on Business

- Accelerate – *Speed up response time* to meet today's instant gratification culture
- Simplify – *Avoid complexity* in communication to not over-tax customer bandwidth
- Multitasking – Let customers accomplish *two goals at once*, i.e. register for event *and* sign up for golf

36

37

Thought Starter
Innovation is part of a ceaseless Darwinian struggle for existence. Of the original 12 Dow Jones companies only one still survives. And that company was just removed from the Dow Jones index. To ensure success, practice relentless innovation.

38

ARTHUR ANDERSEN
BEAR STEARNS
bebe
BLOCKBUSTER
InfoSpace
BORDERS
EASTERN
myspace
GYMBOREE
RadioShack
PAN AM
Payless Shoe Source
RCA
The Learning Company
THE LIMITED
TWA
TOWER RECORDS
Toys R Us
TRUE RELIGION
SPORTS AUTHORITY
Wet Seal
WOOLWORTHS

39

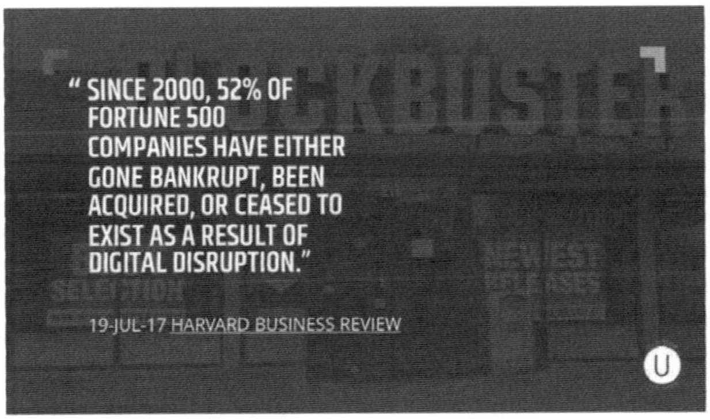
" SINCE 2000, 52% OF FORTUNE 500 COMPANIES HAVE EITHER GONE BANKRUPT, BEEN ACQUIRED, OR CEASED TO EXIST AS A RESULT OF DIGITAL DISRUPTION."
19-JUL-17 HARVARD BUSINESS REVIEW

40

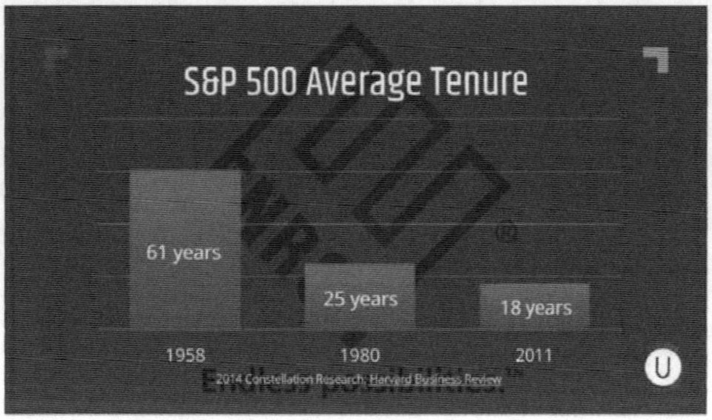

41

"At the present rate of churn, three-quarters of today's S&P 500 will be replaced by 2027."

Jul-17 Innosight; Harvard Business Review

42

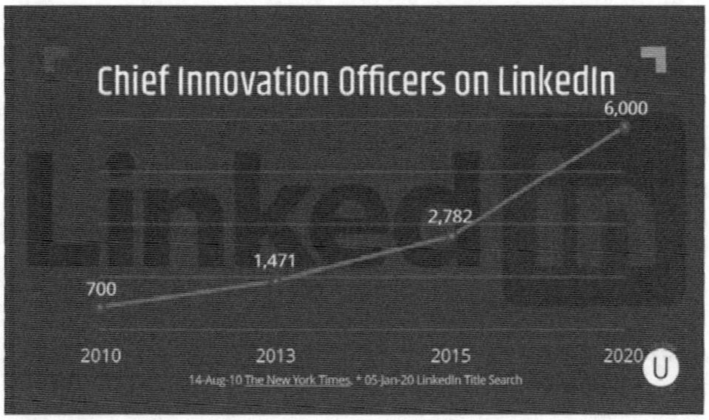

43

THINK ORTHOGONALLY

44

"Think different" ad

Ad campaign reinforces Apple's way of thinking

45

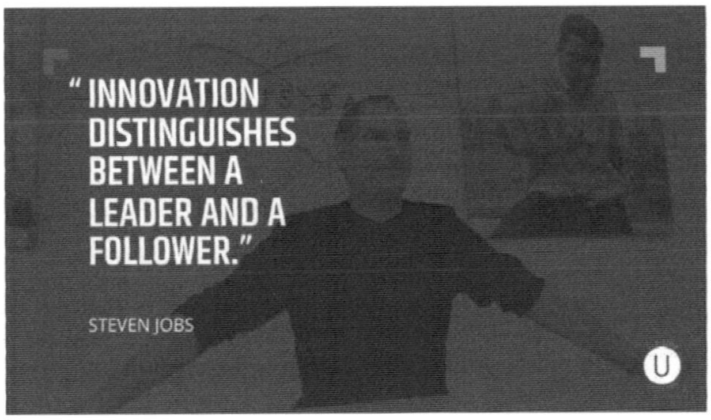

46

"It was a risky proposition said *The New York Times*, calling it 'an aggressive move during an economic slowdown.'"

Michael Tchong, **Ubertrends**

47

"Today, Apple operates **506 stores** in 24 countries [and] generates **nearly $25 billion in sales from its** stores or about 10% of its total annual revenues."

Michael Tchong, **Ubertrends**

48

49

50

"Starbucks will offer free Wi-Fi nationwide, with no registration or account required. Even better, Wi-Fi will not be time limited."

14-Jun-10 Mashable

51

52

"In fiscal 2019, revenues led by 6% comp growth in the U.S. while active U.S. Starbucks Rewards membership was up 15% year-over-year to 17.6 Million."

30-Oct-19 Starbucks/BusinessWire

53

"Loyalty, technology and innovation are continuing to fuel our digital flywheel and propel our business forward."

21-Apr-16 USA Today

54

Top Global Companies by Market Cap

1/1/2010 Rank	Company	Market Cap (B)	1/1/2020 Rank	Company	Market Cap (B)
1.	PetroChina Co. Ltd.	$353	1.	Apple Inc.	$1,305
2.	General Electric Co.	$324	2.	Microsoft Corp.	$1,203
3.	Microsoft Corp.	$271	3.	Alphabet Inc.	$923
4.	ICBC	$269	4.	Amazon.com Inc.	$916
5.	Walmart Inc.	$204	5.	Alibaba Group	$615

15-Jan-20 WPP Investor Day

55

$1.4 TRILLION

56

Twitter Insight

German economist shares unvarnished truth

Katharina Utermöhl
@Economist_Kat

The fact that Germany's 30 largest companies have been overtaken by one single American giant is more than just a statistical quirk — it is a striking example of how Europe's largest economy risks being left behind by the 21st century tech boom ft.com/content/6f6943...

The Apple effect: Germany fears being left behind by Big Tech
As a tech boom threatens to pass the country by, the iPhone maker is worth more than Germany's 30 leading companies
ft.com

57

"Belgian zinc producer EverZinc has set up a dedicated innovation unit and accelerator to develop new zinc-based products."

31-Oct-2018 ee News Power Management

58

"EverZinc Digital plans to focus its activities on the development of new rechargeable energy storage solutions, extended-life packaging and coatings."

31-Oct-2018 ee News Power Management

59

"Zinc8 Energy Solutions developed a patented zinc-air energy storage system that stores energy in the form of zinc particles and contains none of the traditional battery commodities such as lithium, vanadium or cobalt."

17-Jan-20 Renewable Energy World

60

"The zinc industry in Brazil has been commercializing zinc alloys, perceived as higher added value products, for several applications."

July/August 2017 Brazilian Business Review

61

"The Peruvian Ministry of Production wants to increase investment in innovation by ten-fold over the next decade."

10-Aug-17 TechCrunch

62

"Total [Peruvian] venture capital investment will likely exceed $25 million in 2020, doubling what was achieved in 2019."

12-Feb-20 TechCrunch

63

Hub Criativo
do Beato
U

64

Hub Criativo
do Beato
U

65

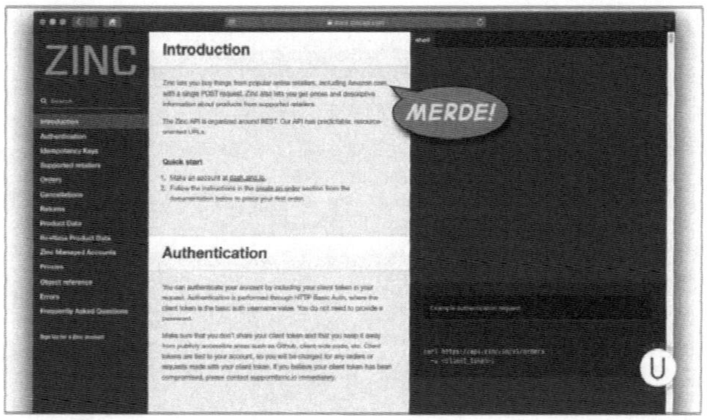
ZINC
Introduction
MERDE!
Quick start
Authentication
U

66

67

Who's familiar with the Quandl API for Commodity Data?

68

http://www.quandl.com/api/v3/datasets/ODA/PZINC_USD

69

AI-Curated Programmatic Buying Platform

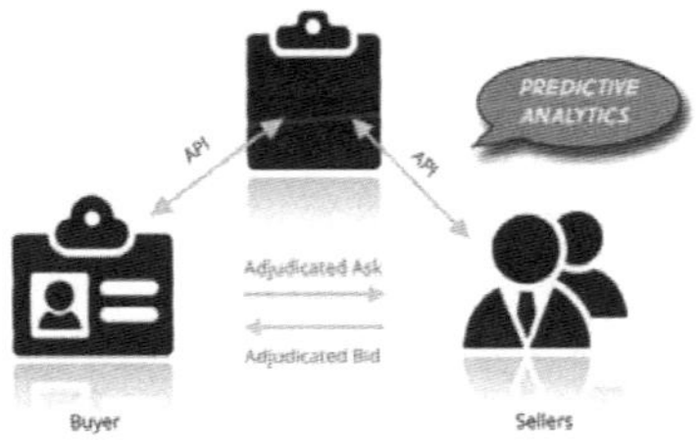

70

"Peru is home to 120 fintech startups actively tackling the issues of financial inclusion and better servicing the region's small and medium-sized businesses."

12-Feb-20 TechCrunch

71

"Shipping giant Maersk blockchain-based TradeLens, a system for tracking customs documentation on goods shipped internationally, registers 10 million shipping events each week."

11-Feb-20 BBC News

72

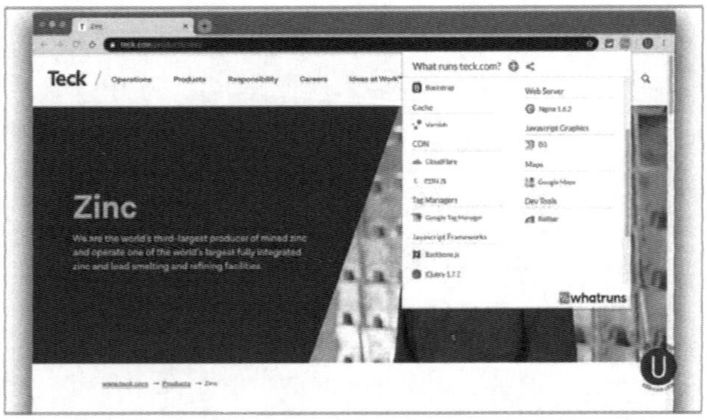

73

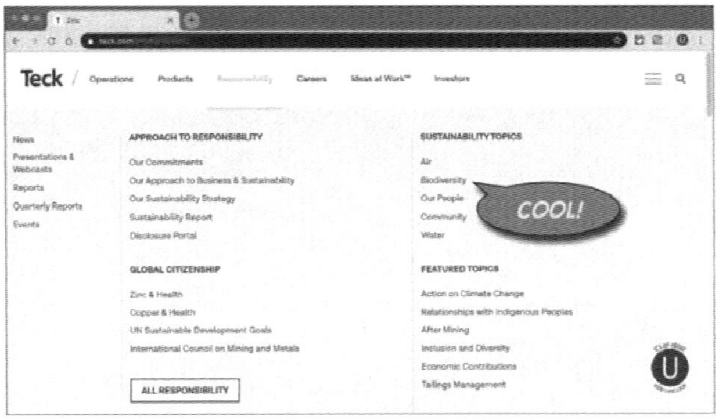

74

What type of Innovation?

Building a **Culture of Innovation**

75

INCREMENTAL INNOVATION Industrial Robots

ORTHOGONAL INNOVATION Toyota Prius

DISRUPTIVE INNOVATION iPhone

76

77

78

Create an Innovation Team

Building a **Culture of Innovation**

79

Gain Knowledge
Building a Culture of Innovation

80

Leverage Resources
Building a Culture of Innovation

81

Build Relationships
Building a Culture of Innovation

82

Use Innovative Tools

Building a **Culture of Innovation**

83

"Surprisingly, according to one study, **administrative innovation** appears to have the greatest impact on company sales."

Michael Tchong, **Ubertrends**

84

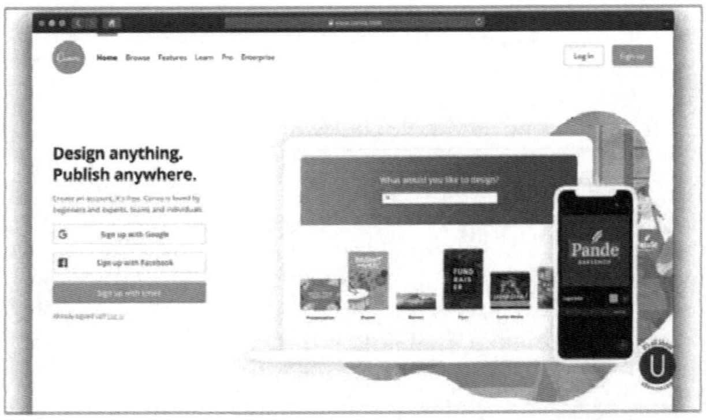

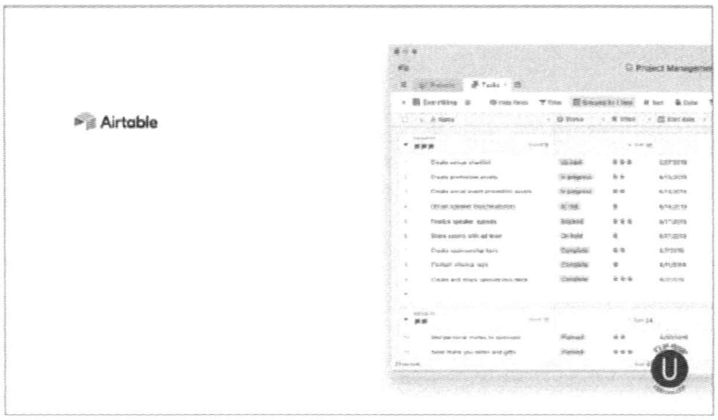
Airtable

85

"IBM picks Slack over Microsoft Teams for its 350,000 employees."
10-Feb-20 The Verge

86

How to create a
culture of
innovation in 2020

87

88

YOUTUBE.COM/C/MICHAELTCHONG

89

"Only 11% of marketers set aside a specific budget for marketing innovation efforts."

Michael Tchong, **Ubertrends**

90

#youcallthatinnovation

91

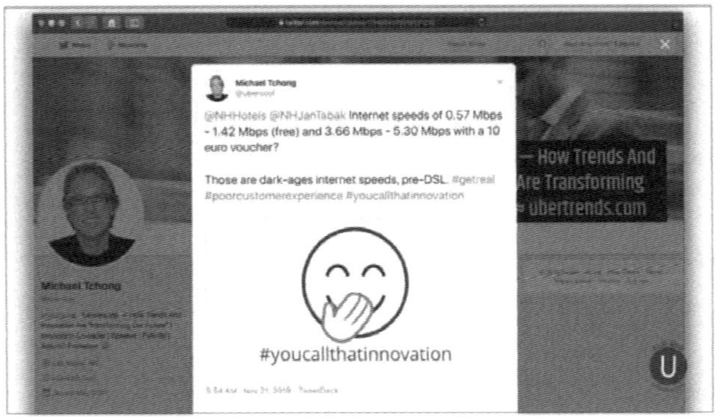

92

Join my Sphere of Influence

Facebook, Instagram, LinkedIn, Twitter: @ubercool

mt@ubercool.com

800-WAY-COOL

ubercool.com

AUTHOR INDEX

International Zinc Association
2530 Meridian Parkway
Suite 115
Durham, NC 27713
USA

ISBN 978-1-7138-0807-7

9 781713 808077

CURRAN ASSOCIATES INC.
proceedings
.com

3rd International Conference on Agricultural and Biological Sciences (ABS 2017)

IOP Conference Series: Earth and Environmental Science
Volume 77

Qingdao, China
26 - 29 June 2017

IOP
Institute of Physics